PUBLISHERS' NOTE

The series in which this title appears was introduced by the publishers in 1957 and is under the general editorship of Dr. Maurice G. Kendall. It is intended to fill a need which has been evident for some time and is likely to grow—the need for some form of publication at moderate cost which will make accessible to a group of readers specialized studies in statistics or special courses on particular statistical topics. There are numerous cases where, for example, a monograph on some newly developed field would be very useful, but the subject has not reached the stage where a comprehensive book is possible or, again, where a course of study is desired in a domain not covered by text-books but where an exhaustive treatment, even if possible, would be expensive and perhaps too elaborate for the readers' needs.

Considerable attention has been given to the problem of producing these books speedily and economically. Appearing in a cover the design of which will be standard, the contents of each volume will follow a simple, straightforward layout, the text production method adopted being suited to the complexity or otherwise of the subject.

The publishers will be interested in approaches from any authors who have work of importance suitable for the series.

CHARLES GRIFFIN & CO. LTD.

OTHER BOOKS ON MATHEMATICS AND STATISTICS

RANDOM WAVELETS
AND
CYBERNETIC SYSTEMS

ENDERS A. ROBINSON, Ph.D.

Associate Professor of Mathematics
in the University of Wisconsin

BEING NUMBER NINE OF

GRIFFIN'S STATISTICAL
MONOGRAPHS & COURSES

EDITED BY

M. G. KENDALL, M.A., Sc.D.

1962

HAFNER PUBLISHING COMPANY

NEW YORK

First published in 1962

PRINTED IN GREAT BRITAIN BY JOHN WRIGHT & SONS LTD., AT THE STONEBRIDGE PRESS, BRISTOL

PREFACE

This monograph has been written for students and research workers in all branches of cybernetics and especially for those in physics and electrical engineering. One of the most important concepts in the science of communication and automatic control has proved to be the concept of *minimum-phase* which was introduced by Hendrik W. Bode in his classic work *Network Analysis and Feedback Amplifier Design* (1945). In this monograph we introduce a new concept which we call *minimum-delay*, and because this new concept is a generalization of the concept of minimum-phase, this volume will be of particular interest to those who are familiar with Bode's work.

The concept of minimum-phase applies to systems with single inputs and single outputs. A minimum-phase system is a system whose phase displacement is the smallest possible for its gain. For such systems, the phase may be determined from the gain.

In November 1955 I had the pleasant experience of visiting the Bell Telephone Laboratories at Murray Hill, New Jersey. There I had very interesting discussions with Dr. Bode, and others who included Dr. Brockway McMillan (now Assistant Secretary, U.S. Air Force), Dr. Claude E. Shannon (now Donner Professor of Science, M.I.T.), Dr. David Slepian, and Professor John W. Tukey (Princeton), about minimum-phase and its relationship to other mathematical and engineering concepts. At that time, the connection of minimum-phase with the early work of N. Wiener, Y. W. Lee, and others on the determination of phase from gain had been demonstrated in Bode's book (1945). Also the connection with Wiener's theory of prediction and filtering had been made clear in the paper of Bode and Shannon (1950). My Ph.D. thesis at the Massachusetts Institute of Technology (1954) had treated the connection of minimum-phase with the predictive decomposition of H. Wold (1938), and with the problem of determining the component wavelets of a time-series by least-squares.

This monograph brings together the above work, and new research, into one exposition. Moreover, we have extended our results to cover the case of systems with multiple inputs and multiple outputs. To do so, we have abandoned the old concept of minimum-phase, and have introduced the new concept of *minimum-delay*. This new concept includes the concept of minimum-phase in the case of systems with single inputs and single outputs. The minimum-delay concept is more general, however, because it also applies to systems with multiple inputs and multiple outputs. This multiple property of minimum-delay is important in applications because as a rule cybernetic systems have many inputs and many outputs.

On the whole the volume is mathematically self-contained and can be read by undergraduate students in engineering and science. Throughout, the terminology and mathematical conventions of electrical engineering are used.

The first two chapters give an account of operational calculus by means of Fourier and Laplace transforms. Chapter III defines wavelets, systems, and inverse systems, and Chapter IV derives a canonical representation of the H^n-stable systems. Chapter V is concerned with the mathematical properties of all-pass systems, which are also called phase-equalizers in communication engineering. Chapter VI gives a theorem of K. Karhunen expressed in terms of wavelets.

The rest of the book is framed around seven theorems on minimum-delay which constitute the main part of our research. Chapter VII contains the first theorem which demonstrates the extremal properties of minimum-delay systems and wavelets. This theorem is in terms of systems with single inputs and single outputs, and wavelets that are scalar-valued. The other six theorems are in terms of systems with multiple inputs and multiple outputs, and wavelets that are matrix-valued.

The methods of least-squares and two-stage least-squares for linear matrix space are presented in Chapter VIII, together with examples of matrix-valued functions. Chapter IX contains the second and third theorems, which carry over and extend the results of the first theorem to multiple systems and matrix-valued wavelets.

In Chapter X the operational calculus is used to develop the theory of multivariate stationary stochastic processes with continuous time-parameter. In this way we are able to present the stochastic integral at a manipulative level. Thus an intermediate course is taken between engineering books—which usually omit the use of the stochastic integral—and advanced mathematical treatises.

The presentation in Chapter X leads up to the problem of the structural identification of multivariate stationary stochastic processes. This problem is explored and the result is the fourth theorem. This theorem states that a minimum-delay structure, and only a minimum-delay structure, can be determined from the covariance of a multivariate stationary stochastic process. The fifth theorem, at the end of the chapter, characterizes the properties of the components of a minimum-delay structure.

In Chapter XI the Wiener theory of filtering and prediction is extended to multivariate stationary processes. Explicit expressions are given for the transfer functions of the multiple systems that yield the optimal filtering and prediction in the sense of least-squares. Finally the last two minimum-delay theorems are given. The sixth shows the relationship of minimum-delay to least-squares and the seventh shows the relationship of minimum-delay to information theory.

The last chapter treats an example of a non-linear feedback control system.

This book has been enriched by the ideas of Professors Philip M. Anselone, Robert N. Buchal, Vincent C. Rideout, and Calvin Wilcox who participated in a seminar on stochastic processes at the University of Wisconsin in the spring of 1960. Likewise, I have benefited from the help of my students at the University of Wisconsin: Richard Black, Calvin McFarling, Stephen M. Robinson, and D. V. Thampuran, who also took part in the seminar and who are carrying out independent research in stochastic processes.

I have also had the good fortune to discuss related problems with Professor David Durand (M.I.T.), Professor Preston C. Hammer (Wisconsin), Professor Stephen M. Simpson (M.I.T.), Dr. Howard A. Slack (The Pure Oil Co.), Dr. Mark K. Smith (Geophysical Service Inc.), and Professor Norbert Wiener (M.I.T.).

In September 1960 I came to the Statistics Institute of Uppsala University at the kind invitation of Professor Herman Wold. During the academic year 1960–1 I was in day-to-day contact with him, and he generously extended to me the wealth of his experience and at all times provided me with much help and encouragement.

While in Scandinavia I have profited a great deal by discussions with Professor Harald Cramér (Stockholm), Professor Ulf Grenander (Stockholm), Dr. Kari Karhunen (Helsinki) and Dr. Ejnar Lyttkens (Uppsala) and with Professor Henry Helson (Berkeley), Dr. David Kendall (Oxford) and Professor Peter Whittle (Manchester) when they visited Uppsala.

It took two years to write this monograph. My research during 1959–60 was sponsored by the United States Army under Contract No. D.A.-11-022-ORD-2059 at the Mathematics Research Center, U.S. Army, University of Wisconsin, Madison, Wisconsin. This work appeared in MRC Technical Reports No. 164 (September 1960), No. 176 (October 1960), and No. 195 (October 1960). My research during 1960–1 at Uppsala University was done under a research fellowship from the University Research Committee of the University of Wisconsin.

Miss Nioleta Hanson and Mr. Lennart Larsson typed the manuscript, Miss Ingrid Åkerrén prepared the final copy of the manuscript, and Mrs. Anita Uhlin drew the figures.

To all, I give my warm thanks.

E. A. Robinson

Uppsala University,
Uppsala, Sweden
June 1, 1962

CONTENTS

NON-MATHEMATICAL INTRODUCTION

When driving an automobile, a person notes the discrepancy, or error, between his steering and the direction of the road. If the error is to the left, he turns the steering wheel to the right, and vice versa. The block diagram is:

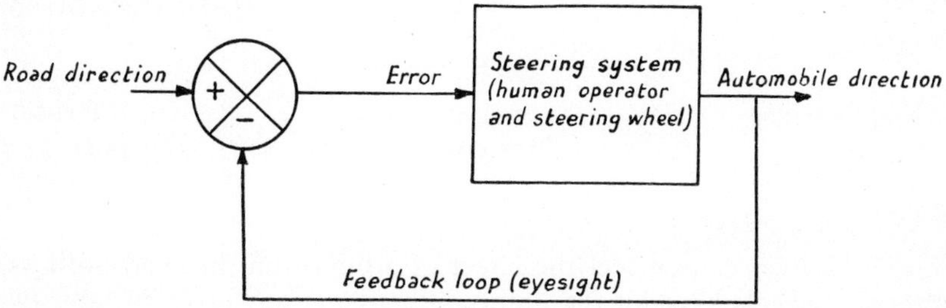

There is always a time-delay between the instant the human operator observes an error and the instant he starts to take corrective action. The operator with the *minimum-delay* is the safest driver.

The desired direction of a ship is set on the gyrocompass. A feedback mechanism indicates the error between the desired direction and the actual direction of the ship. This error activates the guidance system consisting of power amplifiers which force the rudders in the direction that decreases the error. The block diagram is:

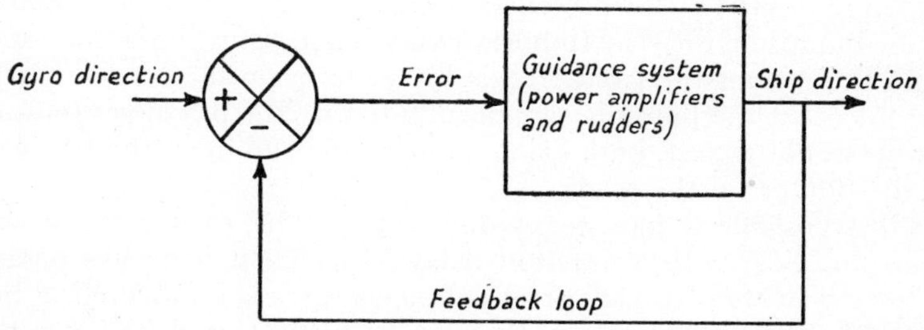

Because it takes time to supply the power for turning the ship, there is a time-delay in the guidance system. Suppose that the ship is off-course to the right. The feedback mechanism indicates an error to the right, and the power amplifiers force the rudders to the left. Because of the time-delay the ship overshoots the gyro direction to the left. Now the feedback mechanism indicates an error to the left, and the power amplifiers force the rudders to the right. Because of the time-delay, the ship again overshoots the gyro direction, this time to the right. The feedback mechanism now indicates an error to the right, and because of the time-delay, a third overshoot is produced, this time to the left. These oscillations about the gyro direction may either increase in magnitude on each successive swing, or decrease. If they increase, the guidance system is unstable; if they decrease, it is stable. Clearly the guidance system with the *minimum-delay* is the one with the greatest possibility of being stable.

An automobile has a system of springs and shock absorbers. The purpose of this spring system is to transmit the ups and downs of a bumpy road into a smoother ride for the passenger. The block diagram is:

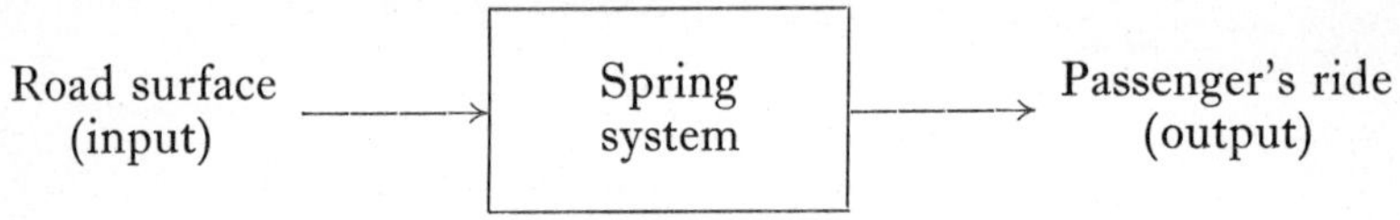

The passenger cannot feel the effect of a bump on the road until or after the wheels have hit the bump. That is, at any given instant the spring system can respond only to inputs that have occurred or are occurring, and cannot respond to inputs which have not yet occurred. Systems that enjoy this property are called *realizable*, and they include all physical systems.

Because of friction in the spring system, the response of the system to any given road bump will eventually die out with time. Systems with this property are called *dissipative*.

The spring system may be described by its *gain* and its *delay*. Its *gain* is a measure of the increase or decrease in the magnitude of the bumps as felt by the passenger (output) compared to the magnitude of the bumps on the road (input). *Delay* is a measure of the time-delay from the instant the front wheels hit a bump on the road (input) to the time that the bump is significantly felt by the passenger (output). As we would expect, both gain and delay depend upon the frequency of the bumps on the road.

It is possible to have many different systems, each with the same gain, but each with a different delay. In fact, it is always possible to have systems with very great delays, as there is no theoretical limit to the greatness of the delay that can be incorporated into a system. On the other hand, there is a limit to the smallness of the delay that a system can possess. The reason is that it always takes some time for a system to respond significantly to an input. The system with the smallest possible delay for its gain is called a *minimum-delay system*.

In the above examples we have decribed some specific systems to illustrate certain concepts and descriptive measures. For the rest of this introductory section, we shall speak about realizable systems in general.

Suppose we have two systems connected in tandem:

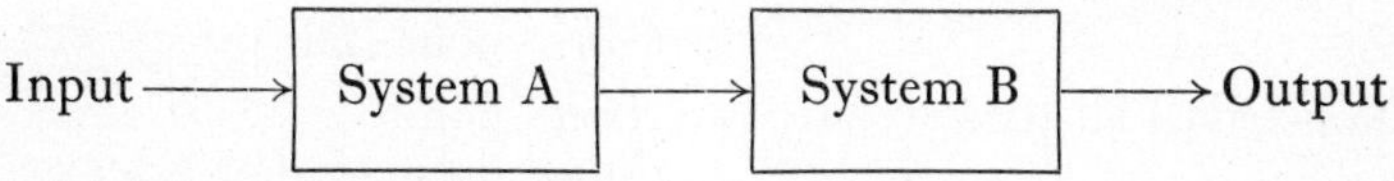

The gain of the overall system is equal to the product of the gains of the component systems, whereas the delay of the overall system is equal to the sum of the delays of the component systems.

Instead of considering the gain, we may consider the *attenuation*, which is defined to be the logarithm of the gain. The logarithm of a product is equal to the sum of the logarithms of the individual factors. Hence the attenuation of the overall system is equal to the sum of the attenuations of the component systems.

In summary, then, for the above diagram, we have

$$\text{Attenuation of overall system} = \text{Attenuation of A} + \text{Attenuation of B},$$

$$\text{Delay of overall system} = \text{Delay of A} + \text{Delay of B}.$$

As we have seen, a realizable system can be described by its attenuation and its delay. Metaphorically, let us think of the attenuation of a system as an "asset" and the delay of the system as the "cost" of this asset. That is, attenuation must always be paid for in delay. In the chapters ahead of us, we shall see that the price that is paid is always larger than (or equal to) the amount of attenuation received.

The realizable systems for which the price paid is equal to the amount of attenuation received are the minimum-delay systems. Hence the minimum-delay is the fair price for the attenuation.

All the other realizable systems, that is, the non-minimum-delay systems, have delays greater than the minimum. Hence, for such a system the price paid for the attenuation is greater than the fair price, and the extra amount paid is the difference between its delay and the minimum-delay.

There are realizable systems that have no attenuation, and these systems are called *all-pass systems*. There are two kinds of all-pass systems. The first kind are the *trivial all-pass systems*, and they are the ones with no delay. Since a trivial all-pass system has no attenuation and no delay, we pay nothing for nothing, which is fair. The other kind are the *non-trivial all-pass systems*, and they are the ones with delay. Since a non-trivial all-pass system has no attenuation but has delay, we pay something for nothing, which is unfair.

We now summarize:

Minimum-delay system:	*Asset*	attenuation
	Cost	minimum-delay (fair price)

Non-minimum-delay system:	*Asset*	attenuation
	Cost	minimum-delay plus extra delay (unfair price)
Trivial all-pass system:	*Asset*	none
	Cost	none (fair price)
Non-trivial all-pass system:	*Asset*	none
	Cost	delay (unfair price)

If we join two systems together, the attenuation of the overall system is the sum of their attenuations, and the delay of the overall system is the sum of their delays. That is, the asset of the overall system is the sum of the assets of the two systems, and the cost of the overall system is the sum of the costs of the two systems.

From the above asset–cost tables, we may see the following. The overall system resulting from connecting a non-trivial all-pass system to a minimum-delay system is non-minimum-delay. Conversely, any non-minimum-delay system can always be decomposed into two tandem systems: one a non-trivial all-pass system and the other a minimum-delay system. These results comprise the so-called *canonical representation* (cf. Sections **6.1, 9.6**).

What does it mean to pay the fair price for an asset? It means that we can sell the asset, and by so doing come back to our original state. A minimum-delay system is a system for which the fair price (i.e. the minimum-delay) has been paid for the attenuation. Consequently, for each minimum-delay system there is a realizable inverse system. If a signal is the input to a minimum-delay system, then we can recover this signal in its original form by passing the output of the minimum-delay system into the inverse system:

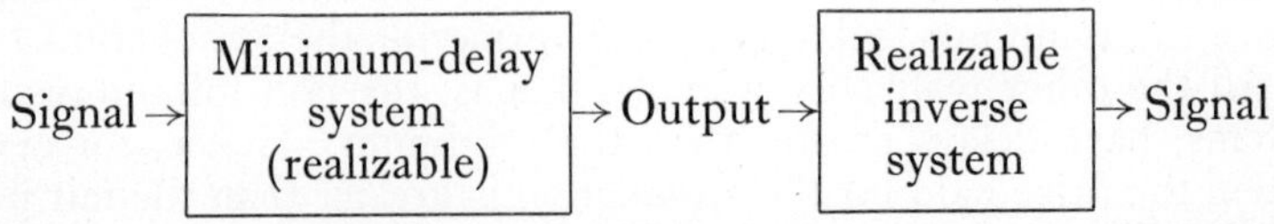

The recovery of the signal is accomplished with no overall time-delay. Hence in the transmission of information from input to output, a minimum-delay system neither destroys nor delays the information.

A non-minimum-delay system is a system for which an unfair price has been paid for the attenuation. Consequently there is no realizable inverse system for a non-minimum-delay system. Some non-minimum-delay systems do not destroy information about the signal, but only delay the information. For these systems, the original signal can be recovered, at least approximately, but with an overall time-delay. Other non-minimum-delay systems (especially multiple ones) destroy information about the signal so that the original signal cannot be recovered, even with an indefinitely long time-delay (cf. Sections **3.4, 7.3, 11.8, 11.9**).

CHAPTER I

TIME-FUNCTIONS

1.1. Stable time-functions

We shall let the real variables t, r, s, τ, and T denote *time*, which is expressed in units of seconds. A function of time, or *time-function*, will be denoted, for example, by $f(t)$. Time-functions may be complex-valued; for example, $f(t) = f_1(t) + if_2(t)$, where $f_1(t)$ and $f_2(t)$ are real-valued functions. We call $f_1(t)$ the *real part* of $f(t)$, denoted by $f_1(t) = \operatorname{Re} f(t)$; and we call $f_2(t)$ the *imaginary part* of $f(t)$, denoted by $f_2(t) = \operatorname{Im} f(t)$.

Definition 1.10. A time-function $f(t)$ is called L^1-*stable* if

$$\int_{-\infty}^{\infty} |f(t)|\, dt < \infty.$$

Definition 1.11. A time-function $f(t)$ is called L^2-*stable* if

$$\int_{-\infty}^{\infty} |f(t)|^2\, dt < \infty.$$

Example 1.10. The time function

$$f(t) = \begin{cases} 0 & t < 1 \\ \dfrac{1}{t} & t \geqslant 1 \end{cases}$$

is L^2-stable but not L^1-stable.

If $f(t)$ is non-zero only over a finite range (t_1, t_2), and if $f(t)$ is L^2-stable, then $f(t)$ is also L^1-stable. If $f(t)$ is non-zero over an infinite range of t, then, as Example 1.10 shows, an L^2-stable $f(t)$ need not be L^1-stable.

Theorem 1.10 (the Schwarz inequality). If $f(t)$ and $g(t)$ are each L^2-stable, then their product $f(t)g(t)$ is L^1-stable, and

$$\int_{-\infty}^{\infty} |f(t)g(t)|\, dt \leqslant \left[\int_{-\infty}^{\infty} |f(t)|^2\, dt \int_{-\infty}^{\infty} |g(s)|^2\, ds\right]^{\frac{1}{2}}.$$

1.2. Convolution

Definition 1.20. The *convolution* of two time-functions $f(t)$ and $g(t)$ is the time function $h(t)$ defined by the integral

$$h(t) = \int_{-\infty}^{\infty} f(t-\tau)g(\tau)\, d\tau.$$

Symbolically, the convolution is denoted by

$$h(t) = f(t) * g(t)$$

or, more concisely, by

$$h = f * g.$$

Theorem 1.20. If $f(t)$ and $g(t)$ each are L^1-stable, then their convolution is L^1-stable.

Proof. By hypothesis, the integrals

$$\int_{-\infty}^{\infty} |g(\tau)| \, d\tau \quad \text{and} \quad \int_{-\infty}^{\infty} |f(t-\tau)| \, dt$$

exist. Hence the integral

$$\int_{-\infty}^{\infty} |g(\tau)| \, d\tau \int_{-\infty}^{\infty} |f(t-\tau)| \, dt$$

exists. Write this integral as

$$\int_{-\infty}^{\infty} dt \int_{-\infty}^{\infty} |g(\tau) f(t-\tau)| \, d\tau.$$

It follows from Fubini's theorem (see Robinson, 1959, p. 24) that

$$\int_{-\infty}^{\infty} |g(\tau) f(t-\tau)| \, d\tau$$

exists, and that

$$\int_{-\infty}^{\infty} dt \left| \int_{-\infty}^{\infty} g(\tau) f(t-\tau) \, d\tau \right|$$

exists. Q.E.D.

Theorem 1.21. Convolution is *commutative*; that is,

$$f * g = g * f.$$

Proof. Let $s = t - \tau$. Then

$$f * g = \int_{-\infty}^{\infty} f(t-\tau) g(\tau) \, d\tau = \int_{-\infty}^{\infty} f(s) g(t-s) \, ds$$

$$= \int_{-\infty}^{\infty} g(t-s) f(s) \, ds = g * f. \qquad \text{Q.E.D.}$$

Theorem 1.22. Convolution is *associative*; that is,

$$(f * g) * h = f * (g * h).$$

Proof. Let $a(t) = f * g$, and $b(t) = g * h$. That is,

$$a(t) = \int_{-\infty}^{\infty} f(t-s) g(s) \, ds,$$

$$b(t) = \int_{-\infty}^{\infty} g(t-\tau) h(\tau) \, d\tau.$$

Then, we have

$$(f*g)*h = a*h = \int_{-\infty}^{\infty} a(t-\tau)\,h(\tau)\,d\tau$$

$$= \int_{-\infty}^{\infty} \left[\int_{-\infty}^{\infty} f(t-\tau-s)\,g(s)\,ds\right] h(\tau)\,d\tau.$$

Let $r = \tau + s$. Then

$$(f*g)*h = \int_{-\infty}^{\infty} \left[\int_{-\infty}^{\infty} f(t-r)\,g(r-\tau)\,dr\right] h(\tau)\,d\tau$$

$$= \int_{-\infty}^{\infty} d\tau \int_{-\infty}^{\infty} dr\, f(t-r)\,g(r-\tau)\,h(\tau)$$

$$= \int_{-\infty}^{\infty} dr \int_{-\infty}^{\infty} d\tau\, f(t-r)\,g(r-\tau)\,h(\tau)$$

$$= \int_{-\infty}^{\infty} dr\, f(t-r)\left[\int_{-\infty}^{\infty} g(r-\tau)\,h(\tau)\,d\tau\right]$$

$$= \int_{-\infty}^{\infty} f(t-\tau)\,b(r)\,dr = f*b$$

$$= f*(g*h).$$

Q.E.D.

Theorem 1.23. Convolution is *distributive* with respect to addition; that is,

$$f*(g+h) = (f*g)+(f*h).$$

Proof.

$$f*(g+h) = \int_{-\infty}^{\infty} f(t-\tau)\,[g(\tau)+h(\tau)]\,d\tau$$

$$= \int_{-\infty}^{\infty} f(t-\tau)\,g(\tau)\,d\tau + \int_{-\infty}^{\infty} f(t-\tau)\,h(\tau)\,d\tau$$

$$= (f*g)+(f*h).$$

Q.E.D.

1.3. The unit-step function

Definition 1.30. The *unit-step function* $u(t)$ is defined by

$$u(t) = 0 \quad \text{for} \quad t<0,$$

$$u(t) = 1 \quad \text{for} \quad t>0.$$

The unit-step function was introduced by Oliver Heaviside. This function has the constant value zero for negative times, has a step of magnitude one at zero, and has the constant value one for positive times. That is, $u(t)$ is continuous for all t, except $t = 0$, where it has a finite discontinuity. The value of $u(t)$ for $t = 0$ need not be defined, or may be defined to be 0, or 0·5, or 1, or some other finite number. We see that $u(t)$ is neither L^1-stable nor L^2-stable.

Example 1.30

$$u(t) = \frac{1}{2}\left(1 + \frac{t}{\sqrt{t^2}}\right).$$

Example 1.31

$$u(t) = \lim_{\epsilon \to 0+} \left[\frac{1}{2} + \frac{1}{\pi}\tan^{-1}\frac{t}{\epsilon}\right], \qquad \epsilon > 0,$$

where $\tan^{-1}(t/\epsilon)$ is given its principal value

$$-\frac{\pi}{2} \leqslant \tan^{-1}\frac{t}{\epsilon} \leqslant \frac{\pi}{2}.$$

1.4. The impulse function

Definition 1.40. The *impulse function* $\delta(t)$ is defined by

$$\delta(t) = 0 \qquad \text{for} \quad t \neq 0,$$
$$\delta(t) = +\infty \quad \text{for} \quad t = 0,$$

with

$$\int_{-\infty}^{\infty} \delta(t)\, dt = 1.$$

The impulse function was extensively used by P. A. M. Dirac, and so is also called the *Dirac delta function*, although it had been used by mathematicians before him. The impulse function has the constant value zero for all negative and positive times, and has infinite magnitude but unit area at zero time. That is, $\delta(t)$ is continuous for all t, except $t = 0$, where it has an infinite discontinuity. The impulse function $\delta(t)$ is L^1-stable by definition, but it is not L^2-stable. The impulse function may be described as a spike, or knife-sharp impulse, of infinite height at $t = 0$, but with total area equal to one.

Example 1.40. Define

$$\delta(t;\, \epsilon) = \frac{1}{\pi}\frac{\epsilon}{t^2 + \epsilon^2}, \qquad \epsilon > 0.$$

We have

$$\lim_{\epsilon \to 0+} \delta(t;\, \epsilon) = \begin{cases} \lim\limits_{\epsilon \to 0+} \dfrac{\epsilon}{t^2} = 0 & \text{if} \quad t \neq 0 \\ \lim\limits_{\epsilon \to 0+} \dfrac{1}{\epsilon} = \infty & \text{if} \quad t = 0. \end{cases}$$

Furthermore,

$$\int_{-\infty}^{\infty} \delta(t;\, \epsilon)\, dt = \int_{-\infty}^{\infty} \frac{1}{\pi}\frac{\epsilon}{t^2 + \epsilon^2}\, dt = 1$$

independently of the numerical value of ϵ ($\epsilon > 0$). We therefore postulate that

$$\lim_{\epsilon \to 0+} \int_{-\infty}^{\infty} \delta(t;\epsilon)\,dt = \int_{-\infty}^{\infty} \lim_{\epsilon \to 0+} \delta(t;\epsilon)\,dt$$

which is

$$1 = \int_{-\infty}^{\infty} \lim_{\epsilon \to 0+} \delta(t;\epsilon)\,dt.$$

Therefore, we have

$$\delta(t) = \lim_{\epsilon \to 0+} \delta(t;\epsilon) = \lim_{\epsilon \to 0+} \frac{1}{\pi} \frac{\epsilon}{t^2+\epsilon^2}, \quad \epsilon > 0.$$

Example 1.41

$$\delta(t-s) = \lim_{\epsilon \to 0+} \frac{1}{\pi} \frac{\epsilon}{(t-s)^2+\epsilon^2}, \quad \epsilon > 0.$$

Example 1.42

$$\delta(t) = \lim_{\alpha \to \infty} \frac{a}{2} \mathrm{e}^{-|\alpha t|}.$$

Example 1.43

$$\delta(t) = \lim_{\epsilon \to 0+} \frac{u(t)-u(t-\epsilon)}{\epsilon}, \quad \epsilon > 0.$$

Example 1.44. From Example 1.31 we see that

$$u(t;\epsilon) = \frac{1}{2} + \frac{1}{\pi} \tan^{-1} \frac{t}{\epsilon}, \quad \epsilon > 0,$$

serves as an approximation for $u(t)$ for small ϵ. The derivative of $u(t;\epsilon)$ is

$$\frac{d}{dt} u(t;\epsilon) = \frac{1}{\pi} \frac{\epsilon}{t^2+\epsilon^2} = \delta(t;\epsilon).$$

(See Example 1.40.) We have

$$\lim_{\epsilon \to 0+} \frac{d}{dt} u(t;\epsilon) = \lim_{\epsilon \to 0+} \delta(t;\epsilon) = \delta(t).$$

We postulate that

$$\lim_{\epsilon \to 0+} \frac{d}{dt} u(t;\epsilon) = \frac{d}{dt} \lim_{\epsilon \to 0+} u(t;\epsilon)$$

which is

$$\delta(t) = \frac{d}{dt} u(t).$$

That is, the impulse function $\delta(t)$ may be considered to be the derivative $u'(t)$ of the unit-step function.

Theorem 1.40. The impulse function is *even*, i.e.

$$\delta(-t) = \delta(t).$$

Theorem 1.41. The impulse function plays the role of *unity* in convolution; that is,

$$f(t) * \delta(t) = \int_{-\infty}^{\infty} f(t-s)\, \delta(s)\, ds = f(t).$$

Proof. We have

$$f(t) * \delta(t) = \int_{-\infty}^{\infty} f(t-\tau)\, \delta(\tau)\, d\tau.$$

But

$$f(t-\tau)\, \delta(\tau) = \left.\begin{cases} 0 & \text{for} \quad \tau \neq 0 \\ f(t)\, \delta(0) & \text{for} \quad \tau = 0 \end{cases}\right\} = f(t)\, \delta(\tau).$$

Thus

$$f(t) * \delta(t) = \int_{-\infty}^{\infty} f(t)\, \delta(\tau)\, d\tau.$$

Since $f(t)$ is a constant with respect to the variable τ in the integration, we have

$$\int_{-\infty}^{\infty} f(t)\, \delta(\tau)\, d\tau = f(t) \int_{-\infty}^{\infty} \delta(\tau)\, d\tau = f(t). \qquad \text{Q.E.D.}$$

In this section and in Section **1.3** we have introduced the unit-step function and the impulse function which are basic in the operational calculus of Oliver Heaviside (English mathematician, born 1850, died 1925). The operational calculus is one of the most important mathematical discoveries, and despite considerable effort spent in the past half-century on its mathematical formalization, it can safely be said that operational calculus has not yet reached its ultimate form. The most promising recent development is the "theory of distributions" developed by L. Schwartz (1950, 1951), although much yet remains to be done.

CHAPTER II

FREQUENCY FUNCTIONS

2.1. The fourier transform

The *fourier transform* is defined by different constructions for different classes of functions. We shall be interested in the transform of a function that belongs to the class of functions that are L^2-stable, and which we shall call the L^2-Fourier transform, or, more briefly, the Fourier transform. Also we shall consider the transform of a function that belongs to the class of functions that are L^1-stable, and which we shall call the L^1-Fourier transform. We shall use the designation "fourier transform" (with a lower-case f) as the generic term.

Throughout, we shall let the real variables ω, λ, μ denote *angular frequency*, which is expressed in units of radians per second. Angular frequency is equal to 2π times cyclical frequency. Cyclical frequency is expressed in units of cycles per second. Whenever we use the word "frequency" we shall always mean angular frequency.

Let us define the complex variable p by

$$p = \sigma + i\omega$$

where σ is a real variable and ω is the real variable denoting frequency. We have

$$\sigma = \operatorname{Re} p, \quad \omega = \operatorname{Im} p$$

where Re stands for "real part of" and Im stands for "imaginary part of". In the p-plane, we see that real frequency ω is measured along the imaginary (or $i\omega$) axis given by $p = i\omega$ (i.e. $\sigma = \operatorname{Re} p = 0$). Accordingly, a *frequency-function* will be denoted, for example, by $F(i\omega)$. Frequency-functions will generally be complex-valued.

Definition 2.10. A frequency-function $\Phi(i\omega)$ is called *L^1-stable* if

$$\frac{1}{2\pi}\int_{-\infty}^{\infty} |\Phi(i\omega)|\, d\omega < \infty.$$

Definition 2.11. A frequency-function $F(i\omega)$ is called *L^2-stable* if

$$\frac{1}{2\pi}\int_{-\infty}^{\infty} |F(i\omega)|^2\, d\omega < \infty.$$

Theorem 2.10 (the Schwarz inequality). If $F(i\omega)$ and $G(i\omega)$ are each L^2-stable, then their product $\Phi(i\omega) = F(i\omega)\,G(i\omega)$ is

L^1-stable, and

$$\frac{1}{2\pi}\int_{-\infty}^{\infty}|\Phi(i\omega)|\,d\omega \leqslant \left[\frac{1}{2\pi}\int_{-\infty}^{\infty}|F(i\omega)|^2\,d\omega\,\frac{1}{2\pi}\int_{-\infty}^{\infty}|G(i\lambda)|^2\,d\lambda\right]^{\frac{1}{2}}.$$

Theorem 2.11 (Plancherel's theorem or the L^2-Fourier transform theorem). Let $f(t)$ be a L^2-stable time-function. Then its *Fourier transform*

$$F(i\omega) = \int_{-\infty}^{\infty} f(t)\,e^{-i\omega t}\,dt$$

exists (in the sense of convergence in the mean) and $F(i\omega)$ is an L^2-stable frequency-function. The time-function $f(t)$ may be recovered by the *inverse Fourier transform*

$$f(t) = \frac{1}{2\pi}\int_{-\infty}^{\infty} F(i\omega)\,e^{i\omega t}\,dt$$

(in the sense of convergence in the mean).

Theorem 2.12 (Parseval's theorem). Let $f(t)$ and $g(t)$ each be L^2-stable, with Fourier transforms $F(i\omega)$ and $G(i\omega)$ respectively. Then we have *Parseval's equality*

$$\int_{-\infty}^{\infty} f(t)\overline{g(t)}\,dt = \frac{1}{2\pi}\int_{-\infty}^{\infty} F(i\omega)\overline{G(i\omega)}\,d\omega.$$

Proof. By hypothesis, $f(t)$ and $\overline{g(t)}$ are L^2-stable, so by Theorem 1.10 (Schwarz inequality) $f(t)\overline{g(t)}$ is L^1-stable. By Theorem 2.11 (Plancherel's theorem) $F(i\omega)$ and $\overline{G(i\omega)}$ are each L^2-stable, and hence by Theorem 2.10 (Schwarz inequality) $F(i\omega)\overline{G(i\omega)}$ is L^1-stable. Therefore the following integrals exist:

$$\begin{aligned}\int_{-\infty}^{\infty} f(t)\overline{g(t)}\,dt &= \int_{-\infty}^{\infty} dt\,\overline{g(t)}\,\frac{1}{2\pi}\int_{-\infty}^{\infty} F(i\omega)\,e^{i\omega t}\,d\omega\\ &= \frac{1}{2\pi}\int_{-\infty}^{\infty} d\omega F(i\omega)\int_{-\infty}^{\infty} dt\,\overline{g(t)}\,e^{i\omega t}\\ &= \frac{1}{2\pi}\int_{-\infty}^{\infty} d\omega F(i\omega)\overline{\left[\int_{-\infty}^{\infty} g(t)\,e^{-i\omega t}\,dt\right]}\\ &= \frac{1}{2\pi}\int_{-\infty}^{\infty} F(i\omega)\overline{G(i\omega)}\,d\omega. \end{aligned}$$

Q.E.D.

Letting $f(t) = g(t)$ in Parseval's Theorem 2.12, we have

Theorem 2.13. Let $f(t)$ be L^2-stable, and let $F(i\omega)$ be its Fourier transform. Then we have *Bessel's equality*

$$\int_{-\infty}^{\infty} |f(t)|^2\,dt = \frac{1}{2\pi}\int_{-\infty}^{\infty} |F(i\omega)|^2\,d\omega.$$

Theorem 2.14 (*convolution theorem*). Let $f(t)$ and $h(t)$ each be L^2-stable time-functions with Fourier transforms $F(i\omega)$ and $H(i\omega)$ respectively. Then we have

$$\int_{-\infty}^{\infty} f(t)\, h(t_0 - t)\, dt = \frac{1}{2\pi} \int_{-\infty}^{\infty} F(i\omega) H(i\omega)\, e^{i\omega t_0}\, d\omega.$$

Proof. In Parseval's equality (Theorem 2.12) let

$$g(t) = \overline{h(t_0 - t)}$$

from which it follows

$$\begin{aligned} G(i\omega) &= \int_{-\infty}^{\infty} \overline{h(t_0 - t)}\, e^{-i\omega t}\, dt \\ &= \int_{-\infty}^{\infty} \overline{h(s)}\, e^{-i\omega(t_0 - s)}\, ds \\ &= e^{-i\omega t_0} \overline{\left(\int_{-\infty}^{\infty} h(s)\, e^{-i\omega s}\, ds \right)}, \end{aligned}$$

which gives

$$\overline{G(i\omega)} = e^{i\omega t_0} H(i\omega).$$

Q.E.D.

Theorem 2.15 (*the L^1-Fourier transform theorem*). Let the frequency function $\Phi(i\omega)$ be L^1-stable. Then its *L^1-Fourier transform*

$$\phi(t) = \frac{1}{2\pi} \int_{-\infty}^{\infty} \Phi(i\omega)\, e^{i\omega t}\, d\omega$$

exists and is a continuous and bounded function of t. (Furthermore, the *Riemann–Lebesgue lemma* asserts that $\phi(t) \to 0$ as $t \to \pm\infty$.) Almost everywhere, and in particular for all ω where $\Phi(i\omega)$ is continuous, we have

$$\Phi(i\omega) = \lim_{T\to\infty} \int_{-T}^{T} \left(1 - \frac{|t|}{T}\right) \phi(t)\, e^{-i\omega t}\, dt.$$

The Convolution Theorem 2.14 may now be restated as follows:

Theorem 2.16 (*convolution theorem*). Let $f(t)$ and $g(t)$ each be L^2-stable time-functions and let $F(i\omega)$ and $G(i\omega)$ be their Fourier transforms. Then the convolution $f*g$ is the L^1-Fourier transform of the product $F(i\omega)\, G(i\omega)$; that is,

$$f*g = \frac{1}{2\pi} \int_{-\infty}^{\infty} F(i\omega)\, G(i\omega)\, e^{i\omega t}\, d\omega.$$

2.2. The one-sided Laplace transform

As in the case of fourier transforms, *laplace transforms* are defined by different constructions for different classes of functions. When we use the term "L^2-Laplace transform" or, more briefly, "Laplace transform", we shall mean the version that can be reduced to an L^2-Fourier transform. As a generic term, we shall use laplace transform (with lower-case l).

Definition 2.20. A time-function $f(t)$ is said to be *one-sided* if $f(t) = 0$ for $t < 0$. Otherwise, a time-function $f(t)$ is said to be *two-sided.*

Clearly, $f(t)$ is one-sided if and only if $f(t) = f(t)u(t)$ where $u(t)$ is the unit-step function: $u(t) = 0$, $t < 0$, and $u(t) = 1$, $t \geqslant 0$.

Definition 2.21. A function $\Phi(p)$ analytic for $\mathrm{Re}\,p = \sigma > 0$ is said to be H^1-*stable* if

$$\frac{1}{2\pi}\int_{-\infty}^{\infty} |\Phi(\sigma + i\omega)|\, d\omega < C \quad \text{for all} \quad \sigma > 0$$

where C is a positive constant that depends only upon the choice of the function Φ, but not on σ.

Definition 2.22. A function $F(p)$ analytic for $\mathrm{Re}\,p = \sigma > 0$ is said to be H^2-*stable* if

$$\frac{1}{2\pi}\int_{-\infty}^{\infty} |F(\sigma + i\omega)|^2\, d\omega < C \quad \text{for all} \quad \sigma > 0$$

where C is a positive constant that depends only upon the choice of the function F, but not on σ.

Theorem 2.20 (the Schwarz inequality). If $F(p)$ and $G(p)$ are each H^2-stable, then their product $\Phi(p) = F(p)G(p)$ is H^1-stable.

Theorem 2.21 (one-sided L^2-Laplace transform theorem). Let $f(t)$ be a one-sided, L^2-stable time-function. Then its *one-sided Laplace transform*

$$F(p) = \int_0^{\infty} f(t)\, \mathrm{e}^{-pt}\, dt$$

exists (in the sense of convergence in the mean) for $\mathrm{Re}\,p = \sigma \geqslant 0$, $F(p)$ is analytic for $\mathrm{Re}\,p = \sigma > 0$, and $F(p)$ is an H^2-stable function.

Proof. Since $p = \sigma + i\omega$, we have

$$F(p) = F(\sigma + i\omega) = \int_0^{\infty} f(t)\, \mathrm{e}^{-\sigma t - i\omega t}\, dt$$

$$= \int_{-\infty}^{\infty} [f(t)\, u(t)\, \mathrm{e}^{-\sigma t}]\, \mathrm{e}^{-i\omega t}\, dt$$

so $F(p)$ is the Fourier transform of $f(t)\,u(t)\,\mathrm{e}^{-\sigma t}$. But $f(t)\,u(t)\,\mathrm{e}^{-\sigma t}$ is L^2-stable for $\sigma \geqslant 0$ because

$$|f(t)\,u(t)\,\mathrm{e}^{-\sigma t}| \leqslant |f(t)\,u(t)| \quad \text{for} \quad \sigma \geqslant 0$$

and so

$$\int_{-\infty}^{\infty} |f(t)\,u(t)\,e^{-\sigma t}|^2\,dt \leqslant \int_0^{\infty} |f(t)|^2\,dt, \quad \sigma \geqslant 0.$$

Therefore, by Plancherel's theorem 2.11, the one-sided Laplace transform $F(p)$ exists for $\sigma \geqslant 0$. That is, $F(\sigma + i\omega)$ considered as a function of $i\omega$ is the Fourier transform of a time-function that is zero for negative t and $e^{-\sigma t} f(t)$ for positive t (where $\sigma \geqslant 0$). Moreover, it can be shown that $F(p)$ is analytic for $\sigma > 0$.

By Theorem 2.13 (Bessel's equality) we have

$$\frac{1}{2\pi}\int_{-\infty}^{\infty} |F(\sigma + i\omega)|^2\,d\omega = \int_0^{\infty} |f(t)\,e^{-\sigma t}|^2\,dt \leqslant \int_0^{\infty} |f(t)|^2\,dt = C$$

for all $\sigma \geqslant 0$, which states that $F(p)$ is H^2-stable. Q.E.D.

Theorem 2.22. If the frequency-function $F(i\omega)$ is L^2-stable, and if its inverse Fourier transform

$$f(t) = \frac{1}{2\pi}\int_{-\infty}^{\infty} F(i\omega)\,e^{i\omega t}\,d\omega$$

is a one-sided time-function, then the Laplace transform

$$F(p) = \int_0^{\infty} f(t)\,e^{-pt}\,dt$$

is analytic for $\operatorname{Re} p > 0$ and H^2-stable.

Proof. Since $F(i\omega)$ is L^2-stable, it follows from Bessel's equality (Theorem 2.13) that $f(t)$ is also L^2-stable. Consequently $f(t)$ is a one-sided, L^2-stable time-function, and Theorem 2.22 reduces to Theorem 2.21. Q.E.D.

The converse to Theorem 2.22 is:

Theorem 2.23. If the function $F(p)$ is analytic for $\operatorname{Re} p > 0$, and if $F(p)$ is an H^2-stable function, then $F(p)$ has the representation

$$F(p) = \int_{-\infty}^{\infty} f(t)\,e^{-pt}\,dt = \int_0^{\infty} f(t)\,e^{-pt}\,dt$$

where $f(t)$ is a one-sided, L^2-stable time-function.

Proof. By hypothesis $F(p)$ is analytic for $\sigma > 0$ and H^2-stable. Hence $F(\sigma + i\omega)$ considered as a function of $i\omega$ (for fixed $\sigma > 0$) is an L^2-stable frequency function. Hence by Plancherel's Theorem 2.11, it is the Fourier transform of an L^2-stable time-function, which we denote by $f(t)\,e^{-\sigma t}$. By Theorem 2.13 (Bessel's equality) we have

$$\int_{-\infty}^{\infty} |f(t)\,e^{-\sigma t}|^2\,dt = \frac{1}{2\pi}\int_{-\infty}^{\infty} |F(\sigma + i\omega)|^2\,d\omega, \quad \sigma > 0.$$

Because $F(p)$ is H^2-stable, the above expression (for $\sigma>0$) is bounded by a positive constant C. That is, we have the inequality

$$\int_{-\infty}^{0}|f(t)|^2 e^{-2\sigma t}\,dt+\int_{0}^{\infty}|f(t)|^2 e^{-2\sigma t}\,dt\leqslant C,\quad \sigma>0, \qquad (2.20)$$

and this inequality can only occur when $f(t)$ vanishes for negative t. Otherwise there will be some interval (t_1, t_2) with $t_1<t_2<0$ over which

$$\int_{t_1}^{t_2}|f(t)|^2\,dt = C_0>0$$

where C_0 is a positive constant. Then we would have

$$\int_{-\infty}^{0}|f(t)|^2 e^{-2\sigma t}\geqslant e^{-2\sigma t_2}C_0$$

which would violate the inequality (2.20) for large enough $\sigma>0$ since $t_2<0$. Q.E.D.

We now wish to introduce the concept of "abscissa of convergence". Let $f(t)$ be a one-sided, L^2-stable time-function, and let σ_1 be the *greatest lower bound* of σ for which

$$\int_{0}^{\infty}|f(t)\,e^{-\sigma t}|^2\,dt<\infty.$$

Since $f(t)$ is L^2-stable this integral converges for $\sigma=0$, so therefore $\sigma_1\leqslant 0$. The one-sided function $u(t)f(t)\,e^{-\sigma t}$ is L^2-stable for $\sigma>\sigma_1$, and hence its Fourier transform

$$F(\sigma+i\omega)=\int_{0}^{\infty}[f(t)\,e^{-\sigma t}]\,e^{-i\omega t}dt$$

exists for $\sigma>\sigma_1$, or, in other words, the one-sided Laplace transform

$$F(p)=\int_{0}^{\infty}f(t)\,e^{-pt}\,dt$$

exists for $\operatorname{Re}p=\sigma>\sigma_1$. Because $F(p)$ exists for values of p to the right of the vertical straight line $\operatorname{Re}p=\sigma_1$, the abscissa σ_1 is called the *abscissa of convergence* of the one-sided Laplace transform $F(p)$. We see that only the real part σ of p is decisive for the convergence of $F(p)$, except on the line $\operatorname{Re}p=\sigma_1$ where the convergence of $F(p)$ does depend upon the imaginary part of p.

Example 2.20. Let $g(t)=u(t)\,e^{-t}$. The function

$$g(t)\,e^{-\sigma t}=u(t)\,e^{-t}e^{-\sigma t}=u(t)\,e^{-(1+\sigma)t}$$

is seen to be L^2-stable when $1+\sigma$ is positive; i.e. $1+\sigma>0$, which is $\sigma>-1$. Hence the abscissa of convergence is $\sigma_1=-1$, and

$$G(p)=\int_{0}^{\infty}e^{-t}e^{-pt}\,dt=\frac{1}{1+p},\quad \operatorname{Re}p>-1,$$

is the one-sided Laplace transform of $u(t)\,\mathrm{e}^{-t}$. By inspection, we see that

$$G(p) = \frac{1}{1+p}$$

is analytic everywhere except at the pole $p = -1$. Nevertheless, as we shall see in Example 2.23, $1/(1+p)$ is the one-sided Laplace transform of $u(t)\,\mathrm{e}^{-t}$ only for $\mathrm{Re}\,p > -1$.

The L^2-Laplace transform or, more briefly, the Laplace transform, exists for time-functions $g(t)$ that are not necessarily L^2-stable, as the following theorem shows.

Theorem 2.24. Let ρ_1 be a real number. Let $g(t)$ be a one-sided time-function for which

$$\int_0^\infty |g(t)|^2 \,\mathrm{e}^{-2\rho_1 t}\, dt = C_1 < \infty$$

where C_1 is a positive constant. Then its Laplace transform

$$G(p) = \int_0^\infty g(t)\,\mathrm{e}^{-pt}\, dt$$

exists (in the sense of convergence in the mean) for $\mathrm{Re}\,p = \sigma \geqslant \rho_1$, $G(p)$ is analytic for $\mathrm{Re}\,p = \sigma > \rho_1$, and $G(p)$ satisfies

$$\frac{1}{2\pi}\int_{-\infty}^{\infty} |G(\sigma+i\omega)|^2\, d\omega = \int_0^\infty |g(t)|^2\,\mathrm{e}^{-2\sigma t}\, dt \leqslant C_1$$

for $\sigma \geqslant \rho_1$.

Proof. By hypothesis, the function $f(t)$ defined by

$$f(t) \equiv u(t)\,g(t)\,\mathrm{e}^{-\rho_1 t} = g(t)\,\mathrm{e}^{-\rho_1 t}$$

is a one-sided, L^2-stable time-function. Applying Theorem 2.21 to this $f(t)$, we obtain Theorem 2.24. Q.E.D.

As a corollary to Theorem 2.24, we obtain:

Theorem 2.25. Let ρ_2 be a real number. Let $h(t)$ be a time-function for which

$$h(t) = 0 \quad \text{for} \quad t > 0$$

and

$$\int_{-\infty}^{0} |h(t)|^2\,\mathrm{e}^{-2\rho_2 t}\, dt = C_2 < \infty$$

where C_2 is a positive constant. Then the Laplace transform

$$H(p) = \int_{-\infty}^{0} h(t)\,\mathrm{e}^{-pt}\, dt$$

exists (in the sense of convergence in the mean) for $\mathrm{Re}\, p = \sigma \leqslant \rho_2$, $H(p)$ is analytic for $\mathrm{Re}\, p = \sigma < \rho_2$, and $H(p)$ satisfies

$$\frac{1}{2\pi}\int_{-\infty}^{\infty} |H(\sigma + i\omega)|^2 \, d\omega = \int_{-\infty}^{0} |h(t)|^2 \, \mathrm{e}^{-2\sigma t} \, dt \leqslant C_2$$

for $\sigma \leqslant \rho_2$.

Proof. By hypothesis, the function $g(t)$ defined by

$$g(t) = h(-t)$$

and the number ρ_1 defined by $\rho_1 = -\rho_2$ satisfy the hypothesis of Theorem 2.24. Applying Theorem 2.24 to this $g(t)$ and ρ_1, we obtain Theorem 2.25. Q.E.D.

Example 2.21. Let us now continue Example 2.20. We see that the function

$$g(t) = u(t)\, \mathrm{e}^{-t}$$

with $\rho_1 = -1 + \epsilon \ (\epsilon > 0)$ satisfies the hypothesis of Theorem 2.24, since

$$\int_{0}^{\infty} |g(t)|^2 \, \mathrm{e}^{-2(-1+\epsilon)t} \, dt = \int_{0}^{\infty} \mathrm{e}^{-2\epsilon t} \, dt = \frac{1}{2\epsilon} = C_1$$

is finite. Hence the Laplace transform of $g(t)$ is

$$G(p) = \frac{1}{1+p}$$

for $\sigma \geqslant \rho_1 = -1 + \epsilon$, where $\epsilon > 0$. The abscissa of convergence, as we have seen in Example 2.20, is

$$\sigma_1 = \text{g.l.b. } \rho_1 = \lim_{\epsilon \to 0} \rho_1 = -1$$

(where g.l.b. stands for "greatest lower bound").

Example 2.22. Let $h(t)$ be defined to be

$$h(t) = -u(-t)\, \mathrm{e}^{-t}$$

where $u(-t)$ is one for negative t, and zero for positive t. With $\rho_2 = -1 - \epsilon \ (\epsilon > 0)$, the function $h(t)$ satisfies the hypothesis of Theorem 2.25, since

$$\int_{-\infty}^{0} |h(t)|^2 \, \mathrm{e}^{-2(-1-\epsilon)t} \, dt = \int_{-\infty}^{0} \mathrm{e}^{2\epsilon t} \, dt = \frac{1}{2\epsilon} = C_2$$

is finite. Hence the Laplace transform of $h(t)$ is

$$H(p) = \int_{-\infty}^{0} -\mathrm{e}^{-t}\, \mathrm{e}^{-pt} \, dt = -\int_{-\infty}^{0} \mathrm{e}^{-(1+p)t} \, dt = \frac{1}{1+p}$$

for $\sigma \leqslant \rho_2 = -1-\epsilon$, where $\epsilon > 0$. The abscissa of convergence σ_2 is

$$\sigma_2 = \text{l.u.b.}\ \rho_2 = \lim_{\epsilon \to 0} \rho_2 = -1$$

(where l.u.b. stands for "least upper bound").

Example 2.23. Let

$$F(p) = \frac{1}{1+p}.$$

Then from Example 2.21, we see that $F(p)$ in the region $\text{Re}\,p > -1$ is the Laplace transform of the one-sided function

$$g(t) = u(t)\,\mathrm{e}^{-t}$$

and from Example 2.22, $F(p)$ in the region $\text{Re}\,p < -1$ is the Laplace transform of the function

$$h(t) = -u(-t)\,\mathrm{e}^{-t}.$$

Since the region $\text{Re}\,p > -1$ includes the $i\omega$ axis (i.e. $\text{Re}\,p = 0$), $g(t)$ is L^2-stable, and since the region $\text{Re}\,p < -1$ excludes the $i\omega$ axis, $h(t)$ is not L^2-stable.

Corresponding to Theorem 2.22, we have:

Theorem 2.26. If the frequency-function $\Phi(i\omega)$ is L^1-stable, and if its L^1-Fourier transform

$$\phi(t) = \frac{1}{2\pi}\int_{-\infty}^{\infty} \Phi(i\omega)\,\mathrm{e}^{i\omega t}\,d\omega$$

is a one-sided time-function, then

$$\Phi(p) = \int_{0}^{\infty} \phi(t)\,\mathrm{e}^{-pt}\,dt$$

is analytic for $\text{Re}\,p > 0$ and H^1-stable.

Corresponding to Theorem 2.23, we have:

Theorem 2.27. If the function $\Phi(p)$ is analytic for $\text{Re}\,p > 0$, and if $\Phi(p)$ is an H^1-stable function, then $\Phi(p)$ has the representation

$$\Phi(p) = \int_{-\infty}^{\infty} \phi(t)\,\mathrm{e}^{-pt}\,dt = \int_{0}^{\infty} \phi(t)\,\mathrm{e}^{-pt}\,dt$$

where $\phi(t)$ is a one-sided, bounded time-function.

2.3. The Laplace transform

Theorem 2.30 (the Laplace transform theorem). Let ρ_1, ρ_2 be two real numbers such that $\rho_1 < \rho_2$.

Let $f(t)$ be a time-function such that

$$\int_0^\infty |f(t)|^2 e^{-2\rho_1 t}\,dt = C_1 < \infty, \tag{2.30}$$

$$\int_{-\infty}^0 |f(t)|^2 e^{-2\rho_2 t}\,dt = C_2 < \infty, \tag{2.31}$$

where C_1 and C_2 are positive constants. Then its *Laplace transform*

$$F(p) = \int_{-\infty}^\infty f(t)\,e^{-pt}\,dt$$

exists (in the sense of convergence in the mean) for

$$\rho_1 \leqslant \operatorname{Re} p = \sigma \leqslant \rho_2,$$

$F(p)$ is analytic for $\rho_1 < \operatorname{Re} p = \sigma < \rho_2$, and $F(p)$ satisfies

$$\frac{1}{2\pi}\int_{-\infty}^\infty |F(\sigma+i\omega)|^2\,d\omega = \int_{-\infty}^\infty |f(t)|^2 e^{-2\sigma t}\,dt \leqslant C_1 + C_2 \tag{2.32}$$

for $\rho_1 \leqslant \sigma \leqslant \rho_2$.

Proof. The function $g(t)$ defined by $g(t) = u(t)f(t)$ satisfies the hypothesis of Theorem 2.24, and the function $h(t)$ defined by $h(t) = u(-t)f(t)$ satisfies the hypothesis of Theorem 2.25. Applying Theorems 2.24 and 2.25, we obtain Theorem 2.23 because $f(t) = h(t) + g(t)$. Q.E.D.

Let ρ_1, ρ_2 be real numbers and $f(t)$ a time-function such that the hypothesis of Theorem 2.30 is satisfied. Then, the *abscissae of convergence* of $F(p)$ are defined to be σ_1, σ_2, where

$$\sigma_1 = \text{g.l.b. } \rho_1,$$
$$\sigma_2 = \text{l.u.b. } \rho_2.$$

The vertical strip in the p-plane enclosed by the two vertical lines $\operatorname{Re} p = \sigma_1$ and $\operatorname{Re} p = \sigma_2$ is called the *strip of convergence* of $F(p)$.

Theorem 2.31. Let $f(t)$ be a time-function with Laplace transform $F(p)$ whose strip of convergence has abscissae σ_1, σ_2. Then $f(t)$ is L^2-stable if and only if the strip of convergence includes the $i\omega$ axis.

Proof. If the strip includes the $i\omega$ axis, then equation (2.32) holds with $\sigma = 0$, and hence $f(t)$ is L^2-stable. Conversely, if $f(t)$ is L^2-stable, then equations (2.30) and (2.31) hold with $\rho_1 = \rho_2 = 0$, so the strip of convergence includes the line $\sigma = 0$, i.e. the $i\omega$ axis. Q.E.D.

Example 2.30. Suppose

$$f(t) = e^{-|t|}.$$

Let

$$g(t) = u(t)f(t), \quad h(t) = u(-t)f(t).$$

Then

$$H(p) = \int_{-\infty}^{0} e^{t}\, e^{-pt}\, dt = \int_{-\infty}^{0} e^{t(1-\sigma)}\, e^{-i\omega t}\, dt$$

converges if $1-\sigma > 0$, i.e. if $\sigma < 1$, so $\sigma_2 = 1$. That is,

$$H(p) = \frac{1}{1-p}, \quad \operatorname{Re} p < 1.$$

From Example 2.20,

$$G(p) = \frac{1}{1+p}, \quad \operatorname{Re} p > -1.$$

Since

$$f(t) = h(t) + g(t),$$

it follows that

$$F(p) = H(p) + G(p) = \frac{1}{1-p} + \frac{1}{1+p}$$

$$= \frac{2}{1-p^2}, \quad -1 < \operatorname{Re} p < 1.$$

2.4. The inverse Laplace transform

Theorem 2.40 (the inverse Laplace transform theorem). Let $f(t)$, ρ_1, ρ_2 and $F(p)$ be defined as in Theorem 2.30. Then the time-function $f(t)$ may be recovered from $F(p)$ by the *inverse Laplace transform*

$$f(t) = \frac{1}{2\pi i} \int_{\sigma - i\infty}^{\sigma + i\infty} F(p)\, e^{pt}\, dp$$

(in the sense of convergence in the mean) for $\rho_1 \leqslant \sigma \leqslant \rho_2$.

Proof. Because $F(p) = F(\sigma + i\omega)$ is the Fourier transform of $f(t)\, e^{-\sigma t}$ (where $\rho_1 \leqslant \sigma \leqslant \rho_2$) we may recover $f(t)\, e^{-\sigma t}$ by the inverse Fourier transform

$$f(t)\, e^{-\sigma t} = \frac{1}{2\pi} \int_{-\infty}^{\infty} F(\sigma + i\omega)\, e^{i\omega t}\, d\omega, \quad \rho_1 \leqslant \sigma \leqslant \rho_2.$$

This equation gives

$$f(t) = \frac{1}{2\pi} e^{\sigma t} \int_{-\infty}^{\infty} F(\sigma + i\omega)\, e^{i\omega t}\, d\omega$$

$$= \frac{1}{2\pi i} \int_{\sigma - i\infty}^{\sigma + i\infty} F(\sigma + i\omega)\, e^{(\sigma + i\omega)t}\, d(\sigma + i\omega) = \frac{1}{2\pi i} \int_{\sigma - i\infty}^{\sigma + i\infty} F(p)\, e^{pt}\, dp$$

where $\rho_1 \leqslant \sigma \leqslant \rho_2$. We see that the inverse Laplace transform yields $f(t)$ independently of the actual value of σ, as long as σ is in the range $\rho_1 \leqslant \sigma \leqslant \rho_2$. That is, once the strip of convergence of $F(p)$ is known, the path of integration from $\sigma - i\infty$ to $\sigma + i\infty$ can be taken along any vertical line inside the strip, always getting the same $f(t)$. Q.E.D.

2.5. Operational rules and theorems

Theorems 2.30 and 2.40 yield a correspondence between a time-function $f(t)$ and its Laplace transform $F(p)$ in the vertical strip $\rho_1 \leqslant \operatorname{Re} p \leqslant \rho_2$. We shall denote this correspondence by

$$f(t) \leftrightarrow F(p), \quad \rho_1 \leqslant \operatorname{Re} p \leqslant \rho_2.$$

Theorems 2.30 and 2.40 tell us that this correspondence is one-to-one as long as we restrict ourselves to the given vertical strip $\rho_1 \leqslant \operatorname{Re} p \leqslant \rho_2$ in the p-plane. As we have seen in Example 2.23, $F(p) = 1/(1+p)$ defined for p in one strip yields one time-function, whereas the same function $1/(1+p)$ but defined for p in another strip yields a different time-function. It is therefore important to specify the strip for which $F(p)$ is defined.

In general, it is necessary to specify the strip of convergence when a one-to-one correspondence is set up between time functions and laplace transforms.

The following rules and theorems of the *operational calculus* of Oliver Heaviside are set forth in a heuristic way. When using these rules, one must be careful to verify the existence of the laplace transforms involved and to specify their strips of convergence. In our work we are usually interested in the strip containing the $p = i\omega$ axis (i.e. the line $\operatorname{Re} p = 0$).

Rule 2.50. If $f(t)$ has laplace transform $F(p)$, then $cf(t)$ has laplace transform $cF(p)$, where c is a complex constant.

Rule 2.51. If $f(t)$ and $g(t)$ have laplace transforms $F(p)$ and $G(p)$, respectively, then $f(t)+g(t)$ has laplace transform $F(p)+G(p)$.

Rule 2.52. If $f(t)$ has laplace transform $F(p)$, then

$$f(t+c) \leftrightarrow e^{cp} F(p)$$

where c is a *real* constant.

Proof.

$$f(t+c) \leftrightarrow \int_{-\infty}^{\infty} f(t+c)\, e^{-pt}\, dt = e^{cp} \int_{-\infty}^{\infty} f(s)\, e^{-ps}\, ds = e^{cp} F(p). \quad \text{Q.E.D.}$$

Rule 2.53. If $f(t)$ has laplace transform $F(p)$, then

$$e^{-ct} f(t) \leftrightarrow F(p+c)$$

where c is a complex constant.

Proof.

$$e^{-ct} f(t) \leftrightarrow \int_{-\infty}^{\infty} [e^{-ct} f(t)]\, e^{-pt}\, dt = F(p+c). \qquad \text{Q.E.D.}$$

Rule 2.54 (*convolution rule*). If $F(p)$ and $G(p)$ are the laplace transforms of $f(t)$ and $g(t)$, respectively, then the product $F(p)\,G(p)$ is the laplace transform of the convolution $f(t) * g(t)$.

Proof. By Rule 2.52 we have

$$f(t-\tau) \leftrightarrow e^{-p\tau} F(p),$$

so by Rule 2.50,

$$f(t-\tau)\, g(\tau)\, d\tau \leftrightarrow e^{-p\tau} F(p)\, g(\tau)\, d\tau.$$

Hence

$$\int_{-\infty}^{\infty} f(t-\tau)\, g(\tau)\, d\tau \leftrightarrow \int_{-\infty}^{\infty} e^{-p\tau} F(p)\, g(\tau)\, d\tau$$

$$= F(p) \int_{-\infty}^{\infty} g(\tau)\, e^{-p\tau}\, d\tau = F(p)\, G(p). \qquad \text{Q.E.D.}$$

Theorem 2.50. The laplace transform of the unit-step function $u(t)$ is

$$\frac{1}{p}, \quad \text{for} \quad \operatorname{Re} p > 0.$$

Proof. We have

$$F(p) = \int_{-\infty}^{\infty} u(t)\, e^{-pt}\, dt = \int_{0}^{\infty} e^{-pt}\, dt = \frac{1}{p} \quad (\text{provided } \operatorname{Re} p > 0).$$

Q.E.D.

Theorem 2.51. The laplace transform of $-u(-t)$ is

$$\frac{1}{p}, \quad \text{for} \quad \operatorname{Re} p < 0.$$

Proof. Since $-u(-t) = -1,\ t<0$, and $-u(-t) = 0,\ t>0$, we have

$$F(p) = \int_{-\infty}^{\infty} -u(-t)\, e^{-pt}\, dt = -\int_{-\infty}^{0} e^{-pt}\, dt = \frac{1}{p} \quad (\text{provided } \operatorname{Re} p < 0).$$

Q.E.D.

Theorem 2.52. The laplace transform of the impulse function $\delta(t)$ is equal to 1 for all p.

Proof. We have, by Theorems 1.40 and 1.41,

$$F(p) = \int_{-\infty}^{\infty} \delta(t)\, e^{-pt}\, dt = \int_{-\infty}^{\infty} \delta(0-t)\, e^{-pt}\, dt = e^{-p\cdot 0} = 1. \qquad \text{Q.E.D.}$$

Theorem 2.53. If t and s are real variables, then

$$\frac{1}{2\pi}\int_{-\infty}^{\infty} e^{ist}\,ds = \delta(t)$$

(this expression being convergent in the sense of a Cesàro limit).

Proof. Since

$$\int_{-\infty}^{\infty} \delta(t)\,e^{-i\omega t}\,dt = 1,$$

we may invert to obtain

$$\frac{1}{2\pi}\int_{-\infty}^{\infty} 1\cdot e^{i\omega t}\,d\omega = 1. \qquad \text{Q.E.D.}$$

Rule 2.55 (differentiation rule). If $f(t)$ has laplace transform $F(p)$, then

$$\frac{d}{dt}f(t) = f'(t) \leftrightarrow pF(p).$$

Proof. Using rules 2.50, 2.51, and 2.52, we have

$$\frac{f(t+\epsilon)-f(t)}{\epsilon} \leftrightarrow \frac{e^{\epsilon p}-1}{\epsilon}F(p)$$

and letting $\epsilon \to 0$ we obtain the desired result. Q.E.D.

Rule 2.56 (integration rule). If $f(t)$ has laplace transform $F(p)$, then

$$\int_{-\infty}^{t} f(s)\,ds \leftrightarrow \frac{1}{p}F(p)$$

for p within the strip of convergence such that $\operatorname{Re} p > 0$; and

$$\int_{\infty}^{t} f(s)\,ds \leftrightarrow \frac{1}{p}F(p)$$

for p within the strip of convergence such that $\operatorname{Re} p < 0$.

Proof. We have

$$\int_{-\infty}^{t} f(s)\,ds = \int_{-\infty}^{\infty} u(t-s)f(s)\,ds = u(t)*f(t).$$

Because (by Theorem 2.50) $u(t) \leftrightarrow 1/p$ (for $\operatorname{Re} p > 0$), it follows from the Convolution Rule 2.54 that

$$u(t)*f(t) \leftrightarrow \frac{1}{p}F(p), \quad \operatorname{Re} p > 0.$$

We have

$$\int_{\infty}^{t} f(s)\,ds = -\int_{t}^{\infty} f(s)\,ds = -\int_{-\infty}^{\infty} u(s-t)f(s)\,ds = -u(-t)*f(t).$$

Because (by Theorem 2.51) $-u(-t) \leftrightarrow 1/p$ (for $\operatorname{Re} p < 0$), it follows that

$$-u(-t) * f(t) \leftrightarrow \frac{1}{p} F(p), \quad \operatorname{Re} p < 0. \qquad \text{Q.E.D.}$$

Repeated application of the Differentiation Rule 2.55 yields:

Rule 2.57 (*n*th *order differentiation rule*). If $f(t)$ has laplace transform $F(p)$, then

$$\left(\frac{d}{dt}\right)^n f(t) = f^{(n)}(t) \leftrightarrow p^n F(p).$$

Theorem 2.54. The laplace transform of the nth derivative $\delta^{(n)}(t)$ of the impulse function is equal to p^n $(-\infty < \operatorname{Re} p < \infty)$.

Proof. Since $\delta(t) \leftrightarrow 1$, the result follows from Rule 2.57. Q.E.D.

Theorem 2.55 (*the Dirac theorem*). If $\delta^{(n)}(t)$ is the nth derivative of the impulse function, then

$$\begin{aligned} f(t) * \delta^{(n)}(t) &= \int_{-\infty}^{\infty} f(t-s)\, \delta^{(n)}(s)\, ds \\ &= \int_{-\infty}^{\infty} f(t)\, \delta^{(n)}(t-s)\, ds = f^{(n)}(t) = \left(\frac{d}{dt}\right)^n f(t). \end{aligned}$$

Proof. By the Convolution Rule 2.54, and Theorem 2.54,

$$f(t) * \delta^{(n)}(t) \leftrightarrow F(p)\, p^n.$$

Application of Rule 2.57 then yields the desired result. Q.E.D.

CHAPTER III

WAVELETS AND CYBERNETIC SYSTEMS

3.1. Wavelets

Definition 3.10. A one-sided, L^2-stable time function is called a *wavelet.*

By definition, then, a time-function $w(t)$ is a wavelet if and only if

$$w(t) = 0, \quad t < 0,$$

$$\int_0^\infty |w(t)|^2 \, dt < \infty.$$

Definition 3.11. Let $w(t)$ be a wavelet. Its Laplace transform

$$W(p) = \int_0^\infty w(t) \, \mathrm{e}^{-pt} \, dt$$

is called the *transfer function* of the wavelet.

Theorems 2.21 and 2.23 may then be expressed as:

Theorem 3.10. A function $W(p)$ is the transfer function of a wavelet if and only if $W(p)$ is analytic in the right half-plane and H^2-stable.

Example 3.10. Consider the wavelet $w(t) = u(t)\,\mathrm{e}^{\alpha t}$ where $u(t)$ is the unit-step function and $\mathrm{Re}\,\alpha < 0$. Then

$$W(p) = \int_0^\infty \mathrm{e}^{\alpha t} \, \mathrm{e}^{-pt} \, dt = \frac{1}{p-\alpha}, \qquad \mathrm{Re}\,p > \mathrm{Re}\,\alpha.$$

Example 3.11. Consider the wavelet

$$w(t) = u(t) \sum_{j=1}^{N} c_j \, \mathrm{e}^{\alpha_j t}$$

where $\alpha_1, \alpha_2, \ldots, \alpha_N$ are distinct, complex numbers with negative real parts, and $c_1, c_2, \ldots, c_N$ are complex constants. Then

$$W(p) = \sum_{j=1}^{N} \frac{c_j}{p-\alpha_j}, \qquad \mathrm{Re}\,p > \mathrm{Re}\,\alpha_j.$$

It is important to note that the rational function of p given here is equal to the transfer function $W(p)$ provided that $\mathrm{Re}\,p > \mathrm{Re}\,\alpha_j$; that is, provided p lies in the region to the right of all the poles α_j of the rational function.

Definition 3.12. Let $w(t)$ be a wavelet. Its *total energy* is defined to be

$$\int_0^\infty |w(t)|^2\,dt = \frac{1}{2\pi}\int_{-\infty}^\infty |W(i\omega)|^2\,d\omega$$

and its *partial energy* (up to time τ) is defined to be

$$\int_0^\tau |w(t)|^2\,dt.$$

3.2. Cybernetic systems

A *cybernetic system* is any configuration (mathematical, physical, biological, social) which over time admits inputs and yields outputs for the purposes of communication and control. Symbolically, we have:

Inputs → [System] → Outputs.

A cause–effect system is one in which the inputs cause the outputs. That is, a cause–effect system yields an output if and only if the system at some time has been, is, or will be activated by an input. We shall call the inputs "stimuli" and the outputs the "responses" to the stimuli:

Stimuli → [Cause–effect system] → Responses.

We shall consider cause–effect systems that admit only one input time-function $f(t)$ and yield only one output time-function $g(t)$:

$f(t)$ → [Cause–effect system] → $g(t)$.

Let us assume that a system has been at rest up to time t_0; more precisely, that the system has been subjected to no input prior to time t_0, so that

$$f(t) \equiv 0, \quad \text{for} \quad t < t_0.$$

At time t_0 an input $f(t)$ is applied. We assume that the input is non-zero over the time interval $(t_0, t_0+\epsilon)$, $\epsilon > 0$; i.e.

$$\int_{t_0}^{t_0+\epsilon} |f(t)|^2\,dt > 0, \quad \epsilon > 0.$$

The resulting output $g(t)$ is called the *normal response* to the input $f(t)$. The system is said to be *realizable* if the normal response satisfies

$$g(t) = 0 \quad \text{for} \quad t < t_0;$$

otherwise the system is said to be *non-realizable*. That is, a system is realizable if and only if the normal response to a (non-zero) input cannot occur prior to the input. In a non-realizable system, the response can precede in time the stimulus that causes the response.

Definition 3.20. Consider a cause–effect system for which a stimulus $f(t)$ causes the normal response $g(t)$. The system is called *time-invariant* if the stimulus $f(t+s)$ causes the normal response $g(t+s)$ for any real number s.

Definition 3.21. Consider a cause–effect system for which a stimulus $f_1(t)$ causes the normal response $g_1(t)$, and the stimulus $f_2(t)$ causes the normal response $g_2(t)$. The system is called *linear* if the stimulus

$$c_1 f_1(t) + c_2 f_2(t)$$

(where c_1 and c_2 are complex constants) causes the normal response

$$c_1 g_1(t) + c_2 g_2(t).$$

Until Chapter IX, we shall use the term "system" to denote a *cause–effect, time-invariant, linear system that admits a single input time-function* $f(t)$ *and yields a single output time-function* $g(t)$.

Definition 3.22. The normal response of a system to an impulse function input $\delta(t)$ is called the *impulse response*, or *weighting function*, $h(t)$, of the system.

That is, the impulse response $h(t)$ is the output resulting from a knife-sharp spike $\delta(t)$ occurring at time $t = 0$:

$$\delta(t) \rightarrow \boxed{\text{System}} \rightarrow h(t).$$

Theorem 3.20. Let a system have impulse response $h(t)$. The response of the system to the input $f(t)$ is given by the convolution

$$\int_{-\infty}^{\infty} f(s)\, h(t-s)\, ds.$$

Proof. Because the system is time-invariant, the response to the impulse $\delta(t-s)$, which occurs at time s, is $h(t-s)$; i.e.

$$\delta(t-s) \rightarrow \boxed{\text{System}} \rightarrow h(t-s).$$

Because the system is linear, we have

$$c_1\,\delta(t) + c_2\,\delta(t-s) \rightarrow \boxed{\text{System}} \rightarrow c_1\, h(t) + c_2\, h(t-s)$$

where c_1 and c_2 are complex constants. In particular, we have

$$f(s_i)\,\delta(t-s_i)\,\Delta s_i \rightarrow \boxed{\text{System}} \rightarrow f(s_i)\,b(t-s_i)\,\Delta s_i,$$

where $f(s_i)\,\Delta s_i$ is a constant, and hence

$$\sum_{i=1}^{n} f(s_i)\,\delta(t-s_i)\,\Delta s_i \rightarrow \boxed{\text{System}} \rightarrow \sum_{i=1}^{n} f(s_i)\,b(t-s_i)\,\Delta s_i.$$

Passing to the limit, we have

$$\int_{-\infty}^{\infty} f(s)\,\delta(t-s)\,ds \rightarrow \boxed{\text{System}} \rightarrow \int_{-\infty}^{\infty} f(s)\,b(t-s)\,ds$$

which is

$$f(t) * \delta(t) \rightarrow \boxed{\text{System}} \rightarrow f(t) * b(t).$$

But $f(t) * \delta(t) = f(t)$, because $\delta(t)$ plays the role of unity in convolution (Theorem 1.41). Hence we have

$$f(t) \rightarrow \boxed{\text{System}} \rightarrow f(t) * b(t),$$

or

$$\text{Stimulus} = f(t) \rightarrow \boxed{\text{System}} \rightarrow \text{Response} = \int_{-\infty}^{\infty} f(s)\,b(t-s)\,ds$$

which shows that the response is a weighted integral average of the stimulus $f(t)$, the *weighting function* being the impulse response function $b(t)$. In other words, the response to the input $f(t)$ is the convolution of the input $f(t)$ with the weighting function (or impulse response) $b(t)$. Q.E.D.

Definition 3.23. A system is called *dissipative* if its impulse response function $b(t)$ is L^2-stable.

Theorem 3.21. A system is realizable and dissipative (R.D.) if and only if its impulse response is a wavelet.

Proof. Since the impulse response $b(t)$ is the normal response to an impulse applied at time $t = 0$, the system is realizable if and only if $b(t) = 0$ for $t < 0$. By definition, the system is dissipative if and only if $b(t)$ is L^2-stable. Hence the system is realizable and dissipative if and only if $b(t)$ is one-sided and L^2-stable; i.e. if and only if $b(t)$ is a wavelet. Q.E.D.

Because of this theorem, we shall denote the impulse response of a realizable–dissipative (R.D.) system by the wavelet $w(t)$. Theorem 3.21 shows that there is a one-to-one correspondence between realizable–dissipative systems and wavelets, and hence we can use two concepts interchangeably.

3.3. Time phenomena and time operators

We have now seen that a time-function may represent:

(1) a time phenomenon; for example, a wavelet,

or

(2) a time operator; for example, a weighting function of a system,

or

(3) both (1) and (2) above; for example, both a wavelet and the weighting function of a realizable–dissipative system.

In any case, we shall use the following terminology to describe a time-function $f(t)$:

Definition 3.30. The laplace transform

$$F(p) = \int_{-\infty}^{\infty} f(t)\, e^{-pt}\, dt$$

is called the *transfer function.*

Definition 3.31. The fourier transform

$$F(i\omega) = \int_{-\infty}^{\infty} f(t)\, e^{-i\omega t}\, dt$$

is called the *gain-and-phase spectrum.* If $F(i\omega)$ is expressed in polar form

$$F(i\omega) = |F(i\omega)|\, e^{iP(i\omega)}$$

then the (real non-negative) function $|F(i\omega)|$ is called the *gain*, and the (real) function $P(i\omega)$ is called the *phase-shift.* The *attenuation* is defined to be $\log|F(i\omega)|$.

If we look at the inverse fourier transform given by

$$f(t) = \frac{1}{2\pi}\int_{-\infty}^{\infty} F(i\omega)\, e^{i\omega t}\, d\omega$$

we see that $f(t)$ may be considered to be a continuous sum of periodic components $e^{i\omega t}$. Using the polar representation of $F(i\omega)$, we have

$$f(t) = \frac{1}{2\pi}\int_{-\infty}^{\infty} |F(i\omega)|\, e^{i[\omega t+P(i\omega)]}\, d\omega.$$

Thus we see that

$$|F(i\omega)|\frac{d\omega}{2\pi}$$

represents the amplification of all periodic components that are inside the band $(\omega, \omega + d\omega)$ and have a common phase-shift $P(i\omega)$ at time $t = 0$.

Accordingly, the *phase-delay* τ_p is determined by

$$e^{i[\omega t + P(i\omega)]} = 1$$

or

$$\omega t + P(i\omega) = 0,$$

which gives

$$t = \tau_p = -\frac{P(i\omega)}{\omega}, \qquad -\infty < \omega < \infty.$$

The *group-delay* τ_g is determined by

$$\frac{d}{d\omega}[\omega t + P(i\omega)] = 0$$

or

$$t + \frac{dP(i\omega)}{d\omega} = 0,$$

which gives

$$t = \tau_g = -\frac{dP(i\omega)}{d\omega}, \qquad -\infty < \omega < \infty.$$

Since

$$\tau_g = -\frac{dP(i\omega)}{d\omega} = \frac{d[\omega\tau_p]}{d\omega},$$

we have the following relationship between group-delay and phase-delay:

$$\tau_g = \tau_p + \omega\frac{d\tau_p}{d\omega}.$$

3.4. Real time and nominal time

The real (i.e. having no imaginary part) variable t represents time. In applications, the real variable t may represent either *real time* or *nominal time*:

Definition 3.40. Real time t is time existent as a thing having actuality.

Definition 3.41. Nominal time t is time existent in name only.

(We note that the "real" used in *real* variable and *real* time has two different meanings.)

For example, a radar system used to track missiles operates in real time. On the other hand, the time-scale on a graph of recorded data is nominal time.

Consider the system with impulse response $b(t)$. If $f(t)$ is the input to the system, then the resulting output is

$$g(t) = \int_{-\infty}^{\infty} f(s)\, b(t-s)\, ds.$$

Let t denote the *present time*, let s (where $s<t$) denote *past time*, and let s (where $s>t$) denote *future time*. Then $g(t)$ at present time t equals

$$g(t) = g_1(t) + g_2(t),$$

where

$$g_1(t) = \int_{-\infty}^{t} f(s)\, b(t-s)\, ds$$

depends upon the past and present values $f(s)$, $s \leqslant t$, of the input, but not upon future values, and

$$g_2(t) = \int_{t+0}^{\infty} f(s)\, b(t-s)\, ds$$

depends upon the future values $f(s)$, $s>t$, of the input, but not upon present and past values. If the system is realizable (i.e. $b(t) = 0$ for $t<0$), then

$$g(t) = g_1(t);$$

that is, the output depends only upon the present and past of the input.

Clearly, a system must be realizable in order to operate in real time ; that is, the present output cannot depend upon future inputs in any real time system. On the other hand, a system need not be realizable to operate in nominal time. For example, let the input $f(t)$ be known over the entire range of nominal time t: $-\infty < t < \infty$. Then certainly, for a given non-realizable weighting function $b(s)$ the output

$$g(t) = \int_{-\infty}^{\infty} f(s)\, b(t-s)\, ds$$

may be computed for any value of t.

Definition 3.42. A system is called a *pure-delay system* if, for input $f(t)$, the output is $f(t-\alpha)$, where $\alpha \geqslant 0$. The non-negative number α is called the *time-delay*.

That is, the output of a pure-delay system is identical in shape to the input, except that the output is delayed α seconds. If $F(p)$ is the laplace transform of the input $f(t)$, then $F(p)\,e^{-p\alpha}$ is the laplace transform of the output $f(t-\alpha)$. Hence the transfer function of a pure-delay system is

$$e^{-p\alpha}, \quad \alpha \geqslant 0.$$

Let $b(t)$ be the impulse response of a non-realizable system. Then $b(t) \neq 0$ for $t < 0$, but we suppose that for large enough delay $\alpha > 0$:

$$b(t) \approx 0 \quad \text{for} \quad t < -\alpha.$$

Then the overall system made up of the pure-delay ($\alpha > 0$) system hitched in tandem to the $b(t)$ system is *approximately realizable*, because

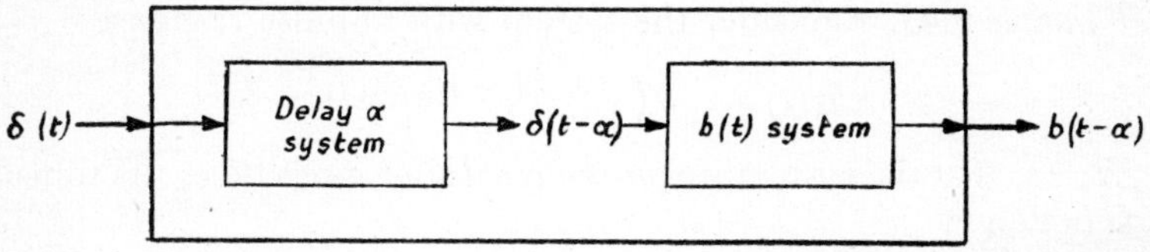

and $b(t-\alpha) \approx 0$ for $t-\alpha < -\alpha$, i.e. for $t < 0$. Hence the non-realizable system can be made approximately realizable by a sufficiently long time-delay α.

In some applications, the objective is to build a system to operate in real time, for example, a radar detection system or a seismic detection system. In other applications, the objective is to construct a mathematical model to simulate a real time phenomenon, such as seismic wave propagation. On the other hand, there are applications in which the objective is to analyse data already completely recorded so that the time scale is nominal time, for example, a seismic record of a blast or earthquake.

Example 3.40. Consider the system with impulse response

$$b_1(t) = u(t)\,e^{\alpha t}, \quad \operatorname{Re}\alpha < 0.$$

We see that the system is *realizable* and *dissipative.* Its transfer function is

$$B_1(p) = \int_0^\infty e^{\alpha t}\, e^{-pt}\, dt = \frac{1}{p-\alpha}, \quad \operatorname{Re} p > \operatorname{Re}\alpha.$$

Example 3.41. Consider the system with impulse response

$$b_2(t) = -u(-t)\,e^{\alpha t}, \quad \operatorname{Re}\alpha < 0.$$

We see that the system is *non-realizable* and *non-dissipative.* Its transfer function is

$$B_2(p) = -\int_{-\infty}^0 e^{\alpha t}\, e^{-pt}\, dt = \frac{1}{p-\alpha}, \quad \operatorname{Re} p < \operatorname{Re}\alpha.$$

In the above two examples, the transfer functions are equal to the same rational function, $1/(p-\alpha)$, but for different regions of the p-plane.

Example 3.42. Consider the system with impulse response

$$b_3(t) = u(t)\,e^{\beta t}, \quad \operatorname{Re}\beta > 0.$$

We see that the system is *realizable* but *non-dissipative.* Its transfer function is

$$B_3(p) = \int_0^\infty e^{\beta t}\,e^{-pt}\,dt = \frac{1}{p-\beta}, \quad \operatorname{Re}p > \operatorname{Re}\beta.$$

Example 3.43. Consider the system with impulse response

$$b_4(t) = -u(-t)\,e^{\beta t}, \quad \operatorname{Re}\beta > 0.$$

We see that the system is *non-realizable* but *dissipative.* Its transfer function is

$$B_4(p) = -\int_{-\infty}^0 e^{\beta t}\,e^{-pt}\,dt = \frac{1}{p-\beta}, \quad \operatorname{Re}p < \operatorname{Re}\beta.$$

In the above two examples, we see that the transfer functions are equal to the same rational function, $1/(p-\beta)$, but for different regions of the p-plane.

3.5. Inverse systems

Definition 3.50. Let the impulse response of a given system be $b(t)$. Then the system with impulse response $b^{-1}(t)$ where

$$b^{-1}(t) * b(t) = \delta(t)$$

is called the *inverse system* to the given system.

By the Convolution Rule 2.54, we have the correspondence:

$$b^{-1}(t) * b(t) = \delta(t) \leftrightarrow B(p)\,B^{-1}(p) = 1$$

where $$b(t) \leftrightarrow B(p), \quad b^{-1}(t) \leftrightarrow B^{-1}(p).$$

Hence the transfer function of the inverse system is given by

$$B^{-1}(p) = \frac{1}{B(p)}.$$

Symbolically, we have

$b(t) \rightarrow$ | Inverse system | $\rightarrow \delta(t)$,

that is, the inverse system contracts the input $b(t)$ into the knife-sharp impulse $\delta(t)$ as output.

Example 3.50. Consider the wavelet $b_1(t)$ given in Example 3.40. The inverse system has transfer function

$$B_1^{-1}(p) = p - \alpha$$

where the strip of convergence includes the $i\omega$ axis, and (using Theorems 2.54 and 2.52) the (realizable) impulse response

$$b_1^{-1}(t) = \delta'(t) - \alpha\delta(t).$$

For verification, we calculate:

$$\begin{aligned} b_1(t) * b_1^{-1}(t) &= b_1'(t) - \alpha b_1(t) = u(t)\,\alpha\,e^{\alpha t} + \delta(t)\,e^{\alpha t} - \alpha u(t)\,e^{\alpha t} \\ &= \delta(t)\,e^{\alpha t} = \delta(t). \end{aligned}$$

Example 3.51. Consider the impulse response $b_4(t)$ in Example 3.43. The inverse system has the (realizable) impulse response

$$b_4^{-1}(t) = \delta'(t) - \beta\delta(t).$$

For verification, we calculate:

$$\begin{aligned} b_4(t) * b_4^{-1}(t) &= b_4'(t) - \beta b_4(t) \\ &= -u(-t)\,\beta\,e^{\beta t} + \delta(t)\,e^{\beta t} + \beta u(-t)\,e^{\beta t} = \delta(t). \end{aligned}$$

Example 3.52. Consider the system with transfer function

$$W_0(p) = \frac{4+p}{(2+p)(3+p)} = \frac{2}{2+p} - \frac{1}{3+p}, \qquad \operatorname{Re} p > -2.$$

By Example 3.11, we have

$$w_0(t) = (2\,e^{-2t} - e^{-3t})\,u(t),$$

which is a wavelet. Differentiating, we have

$$w_0'(t) = (-4\,e^{-2t} + 3\,e^{-3t})\,u(t) + \delta(t).$$

The inverse system has transfer function

$$W_0^{-1}(p) = \frac{(2+p)(3+p)}{4+p} = p + 1 + \frac{2}{p+4}$$

where the strip of convergence includes the $i\omega$ axis, and the (realizable) impulse response

$$w_0^{-1}(t) = \delta'(t) + \delta(t) + 2\,e^{-4t}\,u(t)$$

by Theorem 2.54, Theorem 2.52, and Example 3.10. To verify that $w_0(t) * w_0^{-1}(t) = \delta(t)$, we calculate:

$$\begin{aligned}
w_0(t) * w_0^{-1}(t) &= \int_{-\infty}^{\infty} w_0(\tau)\, w_0^{-1}(t-\tau)\, d\tau \\
&= w_0'(t) + w_0(t) + 2\int_0^t e^{-4(t-\tau)}\, w_0(\tau)\, d\tau \\
&= (-4\,e^{-2t} + 3\,e^{-3t})\, u(t) + \delta(t) + (2\,e^{-2t} - e^{-3t})\, u(t) \\
&\quad + 2\,e^{-4t} \int_0^t e^{4\tau}\, [2\,e^{-2\tau} - e^{-3\tau}]\, d\tau \\
&= \delta(t) + (-2\,e^{-2t} + 2\,e^{-3t})\, u(t) \\
&\quad + 2\,e^{-4t} \left[\int_0^t e^{2\tau}\, 2\, d\tau - \int_0^t e^{\tau}\, d\tau \right] \\
&= \delta(t) + (-2\,e^{-2t} + 2\,e^{-3t})\, u(t) \\
&\quad + 2\,e^{-4t}\, [e^{2t} - 1 - (e^{t} - 1)]\, u(t) \\
&= \delta(t).
\end{aligned}$$

Example 3.53. Consider the system with transfer function

$$\begin{aligned}
W(p) &= \frac{(1-p)(4+p)}{(1+p)(2+p)(3+p)} \\
&= \frac{3}{1+p} - \frac{6}{2+p} + \frac{2}{3+p}, \qquad \operatorname{Re} p > -1.
\end{aligned}$$

By Example 3.11 we have

$$w(t) = (3\,e^{-t} - 6\,e^{-2t} + 2\,e^{-3t})\, u(t),$$

which is a wavelet. The inverse system has transfer function

$$W^{-1}(p) = \frac{(1+p)(2+p)(3+p)}{(1-p)(4+p)} = -p - 3 + 6\frac{3+p}{(1-p)(4+p)}$$

which is

$$W^{-1}(p) = -p - 3 - \frac{6}{5}\frac{1}{p+4} - \frac{24}{5}\frac{1}{p-1},$$

where the strip of convergence includes the $i\omega$ axis. We obtain

$$w^{-1}(t) = -\delta'(t) - 3\delta(t) - \tfrac{6}{5}\, e^{-4t}\, u(t) + \tfrac{24}{5}\, e^{t}\, u(-t).$$

The first three terms in $w^{-1}(t)$ are realizable; the fourth term is not. Nevertheless, the third and fourth terms are each dissipative.

To verify that $w(t) * w^{-1}(t) = \delta(t)$, we calculate:

$$
\begin{aligned}
w(t) * w^{-1}(t) &= \int_{-\infty}^{\infty} w(\tau)\, w^{-1}(t-\tau)\, d\tau \\
&= \int_{-\infty}^{\infty} w(\tau)\, \delta'(t-\tau)\, d\tau - 3\int_{-\infty}^{\infty} w(\tau)\, \delta(t-\tau)\, d\tau \\
&\quad -\frac{6}{5}\int_{0}^{\infty} w(\tau)\, \mathrm{e}^{-4(t-\tau)}\, u(t-\tau)\, d\tau \\
&\quad +\frac{24}{5}\int_{0}^{\infty} w(\tau)\, \mathrm{e}^{t-\tau}\, u(-t+\tau)\, d\tau \\
&= -w'(t) - 3w(t) - \frac{6}{5}\int_{0}^{t} w(\tau)\, \mathrm{e}^{-4(t-\tau)}\, d\tau \\
&\quad +\frac{24}{5}\int_{t}^{\infty} w(\tau)\, \mathrm{e}^{-4(t-\tau)}\, d\tau.
\end{aligned}
$$

Substituting the expression for $w(t)$, and carrying out the calculations, $\delta(t)$ is obtained as expected.

We note that $W_0(p)$ of Example 3.52 and $W(p)$ of Example 3.53 each have the same gain, but that $W_0(p)$ has a realizable inverse whereas $W(p)$ has not.

CHAPTER IV

THE CANONICAL REPRESENTATION

4.1. Definitions and lemmas

In general, a function $F(p)$ of the complex variable $p = \sigma + i\omega$ is called a *transfer function*. The region of the p-plane determined by $\operatorname{Re} p = \sigma > 0$ (with $|p| < \infty$) is called the *right half-plane*. The region of the p-plane determined by $\operatorname{Re} p = \sigma < 0$ (with $|p| < \infty$) is called the *left half-plane*. The boundary of each of these regions is the *iω axis*, determined by $\operatorname{Re} p = \sigma = 0$.

Definition 4.10. A transfer function $F(p)$ that is analytic in the right half-plane is called *H^n-stable* ($n > 0$) if

$$\frac{1}{2\pi} \int_{-\infty}^{\infty} |F(\sigma + i\omega)|^n \, d\omega < C \quad \text{for} \quad \sigma > 0$$

where C is a positive constant that depends upon F and n, but not upon σ.

Definition 4.11. A frequency-function $F(i\omega)$ is called *L^n-stable* ($n > 0$) if

$$\frac{1}{2\pi} \int_{-\infty}^{\infty} |F(i\omega)|^n \, d\omega < \infty.$$

Lemma 4.10. If a transfer function $F(p)$ is analytic in the right half-plane and H^n-stable, then

(a) $F(p)$ tends to a frequency-function $F(i\omega)$ (almost everywhere on the $i\omega$ axis) as $p \to i\omega$,

(b) $F(i\omega)$ is L^n-stable.

Definition 4.12. Let λ be real, and $p = \sigma + i\omega$. Then the *kernel function* $K(p; \lambda)$ is defined to be the rational function

$$K(p; \lambda) = \frac{1}{\pi} \frac{1 - i\lambda p}{p - i\lambda} \frac{1}{1 + \lambda^2}.$$

We note that $K(p; \lambda)$ has a pole at $p = i\lambda$, which is its only singularity. In particular, $K(p; \lambda)$ is analytic in the right half-plane and at the point at infinity.

Lemma 4.11. For fixed λ and σ ($\sigma > 0$) the real part of $K(\sigma + i\omega; \lambda)$ is a *Cauchy density function* with median λ and maximum height $1/\sigma\pi$:

$$\operatorname{Re} K(\sigma + i\omega; \lambda) = \frac{1}{\pi} \frac{\sigma}{\sigma^2 + (\omega - \lambda)^2}.$$

Proof. Let us separate $K(p;\lambda)$ into its real and imaginary parts. Since $p=\sigma+i\omega$, we have

$$K(p;\lambda)=\frac{1}{\pi}\frac{1}{1+\lambda^2}\frac{1-i\lambda p}{p-i\lambda}=\frac{1}{\pi}\frac{1}{1+\lambda^2}\frac{1-i\lambda(\sigma+i\omega)}{\sigma+i\omega-i\lambda}$$

$$=\frac{1}{\pi}\frac{1}{1+\lambda^2}\frac{1+\lambda\omega-i\lambda\sigma}{\sigma+i(\omega-\lambda)}=\frac{1}{\pi}\frac{1}{1+\lambda^2}\frac{[(1+\lambda\omega)-i\lambda\sigma][\sigma-i(\omega-\lambda)]}{\sigma^2+(\omega-\lambda)^2}$$

$$=\frac{1}{\pi}\frac{1}{1+\lambda^2}\frac{[\sigma(1+\lambda\omega)-\lambda\sigma(\omega-\lambda)]-i[(1+\lambda\omega)(\omega-\lambda)+\lambda\sigma^2]}{\sigma^2+(\omega-\lambda)^2}$$

$$=\frac{1}{\pi}\frac{\sigma}{\sigma^2+(\omega-\lambda)^2}-i\frac{1}{\pi(1+\lambda^2)}\frac{(1+\lambda\omega)(\omega-\lambda)+\lambda\sigma^2}{\sigma^2+(\omega-\lambda)^2}.\qquad \text{Q.E.D.}$$

Lemma 4.12.

$$\operatorname{Re}\left[\frac{1-i\lambda p}{p-i\lambda}\right]=\frac{\sigma(1+\lambda^2)}{\sigma^2+(\omega-\lambda)^2}.$$

Lemma 4.13. The real part of $K(p;\lambda)$ tends to the impulse function $\delta(\omega-\lambda)$ as σ tends to zero ($\sigma>0$):

$$\lim_{\sigma\to 0}\operatorname{Re}K(p;\lambda)=\lim_{\sigma\to 0}\frac{1}{\pi}\frac{\sigma}{\sigma^2+(\omega-\lambda)^2}=\delta(\omega-\lambda).$$

Proof. See Section **1.4**. Q.E.D.

Lemma 4.14 (*Jensen's inequality*, Jensen, 1906). If $\log f(\theta)$ is integrable on $-\pi/2\leqslant\theta\leqslant\pi/2$, then

$$\exp\left[\frac{1}{\pi}\int_{-\pi/2}^{\pi/2}\log f(\theta)\,d\theta\right]\leqslant\frac{1}{\pi}\int_{-\pi/2}^{\pi/2}\exp[\log f(\theta)]\,d\theta=\frac{1}{\pi}\int_{-\pi/2}^{\pi/2}f(\theta)\,d\theta.$$

Definition 4.13. Let λ be a real variable. A real, monotone, non-decreasing function $\Gamma(\lambda)$ that is bounded (i.e. $\Gamma(\infty)-\Gamma(-\infty)<\infty$) is called a *distribution function.*

Definition 4.14. A distribution function $\Gamma_c(\lambda)$ is called *absolutely-continuous* if it is the integral of a density function $\gamma(\lambda)$; i.e. if

$$\Gamma_c(\lambda)=\int_{-\infty}^{\lambda}\gamma(\omega)\,d\omega$$

where

$$\gamma(\lambda)\geqslant 0,\qquad \int_{-\infty}^{\infty}\gamma(\omega)\,d\omega<\infty$$

and $\gamma(\lambda)$ has no impulse function components.

Definition 4.15. A distribution function $\Gamma_d(\lambda)$ is called *discrete* if it is a step function with non-negative steps γ_j at λ_j; i.e. if

$$\Gamma_d(\lambda)=\sum_j\gamma_j u(\lambda-\lambda_j)$$

where

$$\gamma_j \geqslant 0, \quad \sum_j \gamma_j < \infty.$$

Since the unit-step function $u(t)$ is the integral of the impulse function $\delta(t)$, we have

$$\Gamma_d(\lambda) = \int_{-\infty}^{\lambda} \sum_j \gamma_j \, \delta(\omega - \lambda_j) \, d\omega = \sum_{\lambda_j \leqslant \lambda} \gamma_j.$$

Lemma 4.15. A distribution function $\Gamma(\lambda)$ may be decomposed into a sum of an absolutely-continuous distribution function $\Gamma_c(\lambda)$ and a discrete distribution function $\Gamma_d(\lambda)$; i.e.

$$\Gamma(\lambda) = \Gamma_c(\lambda) + \Gamma_d(\lambda).$$

(Lebesgue has shown that to make the above Lemma precise, a third component must be added; namely, a continuous but not absolutely-continuous distribution function. We shall omit any further consideration of the continuous, non-absolutely-continuous distribution function, although the reader may easily insert it in what follows if he wishes.)

Definition 4.16. Let λ be a real variable. A real function $\Psi(\lambda)$ is said to have *bounded variation* if

$$\int_{-\infty}^{\infty} |d\Psi(\lambda)| < \infty.$$

Lemma 4.16. A function $\Psi(\lambda)$ of bounded variation can be represented as the difference

$$\Psi(\lambda) = \Gamma(\lambda) - B(\lambda)$$

where $\Gamma(\lambda)$ and $B(\lambda)$ are distribution functions.

Write the distribution function $B(\lambda)$ as the sum of its absolutely-continuous and discrete components:

$$B(\lambda) = B_c(\lambda) + B_d(\lambda) = \int_{-\infty}^{\lambda} \left[\beta(\omega) + \sum_k \beta_k \, \delta(\omega - \lambda_k)\right] d\omega$$

where

$$\beta(\omega) \geqslant 0, \quad \int_{-\infty}^{\infty} \beta(\omega) \, d\omega < \infty, \quad \beta_k \geqslant 0, \quad \sum_k \beta_k < \infty.$$

We then have

Lemma 4.17. A function $\Psi(\lambda)$ of bounded variation has the decomposition:

$$\Psi(\lambda) = \Gamma_c(\lambda) - B_c(\lambda) + \Gamma_d(\lambda) - B_d(\lambda)$$

$$= \int_{-\infty}^{\lambda} \left[\gamma(\omega) - \beta(\omega) + \sum_j \gamma_j \, \delta(\omega - \lambda_j) - \sum_k \beta_k \, \delta(\omega - \lambda_k)\right] d\omega.$$

4.2. H^n-stable transfer functions

Let us consider a transfer function $F(p)$ analytic in the right half-plane and H^n-stable. Let the set $\{p_k\}$ denote the zeros of $F(p)$ that lie in the right half-plane (i.e. $\operatorname{Re} p_k > 0, |p_k| < \infty$).

Example 4.20. Consider the transfer function

$$F(p) = \frac{(1-p)}{(2+p)(3+p)}.$$

$F(p)$ is analytic for $\operatorname{Re} p > 0$ and H^2-stable, and $F(p)$ has the zero $p_1 = 1$ (i.e. $F(1) = 0$) lying in the right half-plane.

If we reflect all the right half-plane zeros p_k about the $i\omega$ axis into the left half-plane, then the function $F_1(p)$ results which has no right half-plane zeros but has the same gain as $F(p)$, as seen in the following example:

Example 4.21. Because $F(p)$, given in Example 4.20, has the zero $p_k = p_1 = 1$ in the right half-plane, we form the function $A_1(p)$ from the zero as follows:

$$A_1(p) = \frac{p_k - p}{\bar{p}_k + p} \frac{|p_k - 1|}{p_k - 1} \frac{|p_k + 1|}{p_k + 1} = \frac{1-p}{1+p}.$$

Since $A_1(p)$ has the same zero as $F(p)$, it follows that $F_1(p)$ given by

$$F_1(p) = \frac{F(p)}{A_1(p)} = \frac{1+p}{1-p} \frac{1-p}{(2+p)(3+p)} = \frac{1+p}{(2+p)(3+p)}$$

is analytic and without zeros in the right half-plane. In other words, the zero $p_1 = 1$ of $F(p)$ has been reflected about the $i\omega$ axis to yield a zero of $F_1(p)$ at $p = -1$. Moreover, $F_1(p)$ is also H^2-stable. We have

$$|A_1(p)| = \left|\frac{1-p}{1+p}\right| = \left[\frac{(1-\sigma)^2 + \omega^2}{(1+\sigma)^2 + \omega^2}\right]^{\frac{1}{2}} < 1$$

for $\operatorname{Re} p = \sigma > 0$ (where $p = \sigma + i\omega$), and hence

$$|F(p)| = |A_1(p)| \, |F_1(p)| < |F_1(p)|$$

in the right half-plane. For $p = i\omega$, we have

$$A_1(i\omega) = \frac{1-i\omega}{1+i\omega} = \exp(-2\tan^{-1}\omega), \quad (-\infty < \omega < \infty),$$

and so $A_1(i\omega)$ has unit gain $|A_1(i\omega)| = 1$ for $-\infty < \omega < \infty$, and phase-shift

$$Q_1(i\omega) = -2\tan^{-1}\omega, \quad (-\infty < \omega < \infty).$$

The phase-delay of $A_1(i\omega)$ is then

$$\tau_p = -\frac{Q_1(i\omega)}{\omega} = \frac{2}{\omega}\tan^{-1}\omega$$

which is positive for all values of ω. Since

$$F(p) = A_1(p) F_1(p),$$

we have

$$|F(i\omega)| = |A_1(i\omega)|\,|F_1(i\omega)| = |F_1(i\omega)|,$$

so $F(i\omega)$ and $F_1(i\omega)$ have the same gain. Because phase-delay is additive, we see that the phase-delay of $F(i\omega)$ is equal to the phase-delay of $A_1(i\omega)$ plus the phase-delay of $F_1(i\omega)$. Since the phase-delay of $A_1(i\omega)$ is positive, it follows that the phase-delay of $F(i\omega)$ is greater than the phase-delay of $F_1(i\omega)$.

Examples 4.20 and 4.21 lead to the following theorems. (Note: A *null* function is a function that vanishes almost everywhere.)

Theorem 4.20. A non-null transfer function $F(p)$
(a) is analytic in the right half-plane,
(b) is H^n-stable,
(c) has the zeros $\{p_k\}$ in the right half-plane,
if and only if

$$F(p) = A_1(p) F_1(p)$$

where

(d) $$A_1(p) = \prod_k \frac{p_k - p}{\bar{p}_k + p} \frac{|p_k - 1|}{p_k - 1} \frac{|p_k + 1|}{p_k + 1}$$

with the $\{p_k\}$ satisfying $\mathrm{Re}\, p_k > 0$, $|p_k| < \infty$, and

$$\sum_k \frac{\mathrm{Re}\, p_k}{1 + |p_k|^2} < \infty,$$

(e) $F_1(p)$ is analytic in the right half-plane, is H^n-stable, and has no zeros in the right half-plane.

Theorem 4.21. Let $F(p)$ and $F_1(p)$ be given as in Theorem 4.20, and let the set $\{p_k\}$ be non-empty. Then, in the right half-plane,

$$|F(p)| < |F_1(p)|;$$

and, almost everywhere on the $i\omega$ axis,

$$|F(i\omega)| = |F_1(i\omega)|.$$

Theorem 4.22. Let $A_1(p)$ be given as in Theorem 4.20(d), and let the set $\{p_k\}$ be non-empty. Then in the right half-plane
(a) $A_1(p)$ is analytic,
(b) $A_1(p)$ has zeros $\{p_k\}$,
(c) $|A_1(p)| < 1$;
and almost everywhere on the $i\omega$ axis

$$|A_1(i\omega)| = 1.$$

4.3. H^n-stable transfer functions without zeros in the right half-plane

Let us consider $F_1(p)$ given as in Theorem 4.20(e); that is, $F_1(p)$ is analytic in the right half-plane, is H^n-stable, and has no zeros in the right half-plane.

Let $e^{-p\alpha}$, $\alpha > 0$, be the transfer function of a pure time delay system (see Definition 3.42). Then the system made up of the delay system in tandem with the $F_1(p)$ system has transfer function

$$e^{-\alpha p} F_1(p), \quad \alpha > 0.$$

This function is a H^n-stable transfer function, analytic and without zeros in the right half-plane, and has the same gain as $F_1(i\omega)$:

$$|e^{-i\omega\alpha} F_1(i\omega)| = |F_1(i\omega)|.$$

Another example of a transfer function that is H^n-stable, analytic and without zeros in the right half-plane, and with the same gain as $F_1(i\omega)$, is

$$e^{-(1/p)} F_1(p).$$

The following theorems give a canonical representation of all H^n-stable transfer functions, analytic and without zeros in the right half-plane, and with the same gain.

Theorem 4.30. A transfer function $F_1(p)$

(a) is analytic in the right half-plane,

(b) is H^n-stable,

(c) has no zeros in the right half-plane,

if and only if

$$F_1(p) = A_0 A_2(p) A_3(p) F_0(p)$$

where

(d) A_0 is a constant with modulus $|A_0| = 1$,

(e) $A_2(p) = e^{-\alpha p}$ with $\alpha \geqslant 0$,

(f) $$A_3(p) = \exp\left[-\frac{1}{\pi}\sum_k \frac{1-i\lambda_k p}{p-i\lambda_k}\beta_k\right]$$

with λ_k real, $\beta_k \geqslant 0$, and $\sum_k \beta_k < \infty$,

(g) $$F_0(p) = \exp\left[\frac{1}{\pi}\int_{-\infty}^{\infty} \frac{1-i\lambda p}{p-i\lambda}\,\frac{\log M(i\lambda)}{1+\lambda^2}\,d\lambda\right]$$

with $M(i\lambda) \geqslant 0$, $M(i\lambda)$ L^n-stable, and

$$\int_{-\infty}^{\infty} \frac{\log M(i\lambda)}{1+\lambda^2}\,d\lambda > -\infty.$$

Proof ("if" statement). By hypothesis,

$$F_1(p) = A_0 A_2(p) A_3(p) F_0(p)$$

where A_0, $A_2(p)$, $A_3(p)$, $F_0(p)$ are given by (d), (e), (f), (g), respectively, in the statement of the Theorem. Clearly, in the right half-plane $A_2(p)$ and $A_3(p)$ are analytic and without zèros.

By hypothesis,

$$\int_{-\infty}^{\infty} \frac{|\log M(i\lambda)|}{1+\lambda^2} d\lambda < \infty.$$

Hence

$$\frac{1}{\pi}\int_{-\infty}^{\infty} \frac{1-i\lambda p}{p-i\lambda} \frac{\log M(i\lambda)}{1+\lambda^2} d\lambda$$

converges and is analytic in the right half-plane, from which it follows that

$$F_0(p) = \exp\left[\frac{1}{\pi}\int_{-\infty}^{\infty} \frac{1-i\lambda p}{p-i\lambda} \frac{\log M(i\lambda)}{1+\lambda^2} d\lambda\right] \tag{4.30}$$

is analytic and without zeros in the right half-plane. Hence $F_1(p)$ is analytic and without zeros in the right half-plane.

Using Definition 4.12, we may write

$$F_0(p) = \exp\left[\int_{-\infty}^{\infty} K(p;\lambda) \log M(i\lambda)\, d\lambda\right].$$

Hence, using Lemma 4.11, we have

$$\begin{aligned}|F_0(p)| &= \exp\left[\operatorname{Re}\int_{-\infty}^{\infty} K(p;\lambda)\log M(i\lambda)\, d\lambda\right] \\ &= \exp\left[\int_{-\infty}^{\infty} [\operatorname{Re} K(p;\lambda)] \log M(i\lambda)\, d\lambda\right] \\ &= \exp\left[\frac{1}{\pi}\int_{-\infty}^{\infty} \frac{\sigma}{\sigma^2+(\omega-\lambda)^2} \log M(i\lambda)\, d\lambda\right].\end{aligned}$$

Define the angle θ by

$$\tan\theta = \frac{\lambda-\omega}{\sigma}, \quad p = \sigma + i\omega, \quad \sigma > 0.$$

Then when $\lambda = -\infty$, $\theta = -\pi/2$; and when $\lambda = \infty$, $\theta = \pi/2$. We have

$$\theta = \tan^{-1}\frac{\lambda-\omega}{\sigma},$$

$$d\theta = \frac{\sigma}{\sigma^2+(\lambda-\omega)^2} d\lambda.$$

Hence

$$|F_0(p)| = \exp\left[\frac{1}{\pi}\int_{-\pi/2}^{\pi/2} \log M(i\lambda)\, d\theta\right],$$

which gives

$$|F_0(p)|^n = \exp\left[\frac{1}{\pi}\int_{-\pi/2}^{\pi/2} n \log M(i\lambda)\, d\theta\right]$$
$$= \exp\left[\frac{1}{\pi}\int_{-\pi/2}^{\pi/2} \log M^n(i\lambda)\, d\theta\right].$$

By Jensen's inequality (Lemma 4.14), we have

$$\exp\left[\frac{1}{\pi}\int_{-\pi/2}^{\pi/2} \log M^n(i\lambda)\, d\theta\right] \leqslant \frac{1}{\pi}\int_{-\pi/2}^{\pi/2} \exp\left[\log M^n(i\lambda)\right] d\theta.$$

Thus we have

$$|F_0(p)|^n \leqslant \frac{1}{\pi}\int_{-\pi/2}^{\pi/2} M^n(i\lambda)\, d\theta = \frac{1}{\pi}\int_{-\infty}^{\infty} M^n(i\lambda)\frac{\sigma}{\sigma^2+(\omega-\lambda)^2}\, d\lambda.$$

Hence, for $\sigma > 0$, we have

$$\frac{1}{2\pi}\int_{-\infty}^{\infty} |F_0(\sigma+i\omega)|^2\, d\omega \leqslant \frac{1}{2\pi}\int_{-\infty}^{\infty} d\omega \int_{-\infty}^{\infty} M^n(i\lambda)\frac{1}{\pi}\frac{\sigma}{\sigma^2+(\omega-\lambda)^2}\, d\lambda$$
$$= \frac{1}{2\pi}\int_{-\infty}^{\infty} M^n(i\lambda)\, d\lambda \int_{-\infty}^{\infty}\frac{1}{\pi}\frac{\sigma}{\sigma^2+(\omega-\lambda)^2}\, d\lambda$$
$$= \frac{1}{2\pi}\int_{-\infty}^{\infty} M^n(i\lambda)\, d\lambda = C.$$

By hypothesis, $M(i\lambda)$ is L^n-stable, and so $C < \infty$. Therefore $F_0(p)$ is H^n-stable. We have, for $\sigma > 0$,

$$|A_2(p)| = |e^{-\alpha p}| = e^{-\alpha\sigma} \leqslant 1, \tag{4.31}$$

$$|A_3(p)| = \exp \operatorname{Re}\left[-\frac{1}{\pi}\sum_k \frac{(1+\lambda_k)\,\omega - i\lambda_k\,\sigma}{\sigma + i(\omega-\lambda_k)}\beta_k\right]$$
$$= \exp\left[-\frac{1}{\pi}\sum_k \frac{\sigma(1+\lambda_k)^2}{\sigma^2+(\omega-\lambda_k)^2}\beta_k\right] \leqslant 1. \tag{4.32}$$

Therefore, we have, for $\sigma > 0$,

$$|F_1(p)| = |A_0|\,|A_2(p)|\,|A_3(p)|\,|F_0(p)| \leqslant |F_0(p)|, \tag{4.33}$$

and hence

$$\frac{1}{2\pi}\int_{-\infty}^{\infty} |F_1(p)|^2\, d\omega \leqslant \frac{1}{2\pi}|F_0(p)|^2\, d\omega \leqslant C,$$

so $F_1(p)$ is H^n-stable. Q.E.D.

Proof ("only if" statement). By hypothesis, $F_1(p)$ is analytic and without zeros in the right half-plane and H^n-stable. Hence $\log F_1(p)$ is analytic in the right half-plane and it can be shown that

$$\int_{-\infty}^{\infty} \frac{\log|F_1(\sigma+i\omega)|}{1+\omega^2}\, d\omega$$

is finite for $\sigma > 0$. We may accordingly represent $\log F_1(p)$ by Green's formula

$$\log F_1(p) = \frac{1}{\pi}\int_{-\infty}^{\infty} \frac{1-i\lambda p}{p-i\lambda}\, d\Psi(\lambda) + ic \tag{4.34}$$

where c is a real constant, and $\Psi(\lambda)$ is a function of bounded variation. The real part of equation (4.34) is (by Lemma 4.12)

$$\begin{aligned} \log|F_1(p)| &= \frac{1}{\pi}\int_{-\infty}^{\infty} \operatorname{Re}\left[\frac{1-i\lambda p}{p-i\lambda}\right] d\Psi(\lambda) \\ &= \frac{1}{\pi}\int_{-\infty}^{\infty} \frac{\sigma}{\sigma^2+(\omega-\lambda)^2}(1+\lambda^2)\, d\Psi(\lambda). \end{aligned} \tag{4.35}$$

By the convolution property of the impulse function, we have

$$\int_{-\infty}^{\infty} \delta(\lambda-\omega)(1+\lambda^2)\, d\Psi(\lambda) = (1+\omega^2)\frac{d\Psi(\omega)}{d\omega}.$$

We may now determine $\Psi(\lambda)$ as follows:

$$\begin{aligned} \Psi(\lambda) &= \int_{-\infty}^{\lambda} d\Psi(\omega) = \int_{-\infty}^{\lambda} \frac{d\omega}{1+\omega^2}(1+\omega^2)\frac{d\Psi(\omega)}{d\omega} \\ &= \int_{-\infty}^{\lambda} \frac{d\omega}{1+\omega^2}\int_{-\infty}^{\infty} \delta(\lambda-\omega)(1+\lambda^2)\, d\Psi(\lambda) \\ &= \int_{-\infty}^{\lambda} \frac{d\omega}{1+\omega^2}\int_{-\infty}^{\infty} \left[\lim_{\sigma\to 0}\frac{1}{\pi}\frac{\sigma}{\sigma^2+(\lambda-\omega)^2}\right](1+\lambda^2)\, d\Psi(\lambda) \\ &= \lim_{\sigma\to 0}\int_{-\infty}^{\lambda} \frac{d\omega}{1+\omega^2}\int_{-\infty}^{\infty} \left[\frac{1}{\pi}\frac{\sigma}{\sigma^2+(\lambda-\omega)^2}(1+\lambda^2)\right] d\Psi(\lambda) \\ &= \lim_{\sigma\to 0}\int_{-\infty}^{\lambda} \frac{d\omega}{1+\omega^2}\int_{-\infty}^{\infty} \operatorname{Re}\left[\frac{1}{\pi}\frac{1-i\lambda p}{p-i\lambda}\right] d\Psi(\lambda) \quad \text{(by Lemma 4.12)} \\ &= \lim_{\sigma\to 0}\int_{-\infty}^{\lambda} \frac{d\omega}{1+\omega^2}\log|F_1(p)| \quad \text{(by equation (4.35))} \\ &= \lim_{\sigma\to 0}\int_{-\infty}^{\lambda} \frac{\log|F_1(\sigma+i\omega)|}{1+\omega^2}\, d\omega. \end{aligned} \tag{4.36}$$

Because $\Psi(\lambda)$ has bounded variation, Lemma 4.17 applies, and equation (4.36) becomes

$$\begin{aligned} \Psi(\lambda) &= \lim_{\sigma\to 0}\int_{-\infty}^{\lambda} \frac{\log|F_1(\sigma+i\omega)|}{1+\omega^2}\, d\omega \\ &= \int_{-\infty}^{\lambda} [\gamma(\omega)-\beta(\omega)+\sum_j \gamma_j\,\delta(\omega-\lambda_j)-\sum_k \beta_k\,\delta(\omega-\lambda_k)]\, d\omega. \end{aligned} \tag{4.37}$$

If $\omega \neq \lambda_j$ or λ_k, then $\delta(\omega-\lambda_j)=\delta(\omega-\lambda_k)=0$, and it follows from equation (4.37) that

$$\gamma(\omega)-\beta(\omega) = \lim_{\sigma\to 0} \frac{\log|F_1(\sigma+i\omega)|}{1+\omega^2} = \frac{\log|F_1(i\omega)|}{1+\omega^2}. \tag{4.38}$$

If we let $\log^+ a$ denote the maximum of $\log a$ and zero, then clearly $\log^+ a \geqslant 0$. If we let $\log^- a$ denote the minimum of $\log a$ and zero, then clearly $\log^- a \leqslant 0$. Here $a \geqslant 0$. Moreover, we have

$$\log a = \log^+ a + \log^- a.$$

Hence we have

$$\Psi(\lambda) = \lim_{\sigma\to 0} \int_{-\infty}^{\lambda} \frac{\log^+|F(\sigma+i\omega)|}{1+\omega^2}\, d\omega + \lim_{\sigma\to 0} \int_{-\infty}^{\lambda} \frac{\log^-|F(\sigma+i\omega)|}{1+\omega^2}\, d\omega$$

where the first term is a bounded, monotone, non-decreasing function of λ, and the second term is a bounded, monotone, non-increasing function of λ. Hence, since

$$\Psi(\lambda) = \Gamma(\lambda) - B(\lambda)$$

where $\Gamma(\lambda)$ is bounded, monotone, non-decreasing and $-B(\lambda)$ is bounded, monotone, non-increasing, we have

$$\Gamma(\lambda) = \lim_{\sigma\to 0} \int_{-\infty}^{\lambda} \frac{\log^+|F(\sigma+i\omega)|}{1+\omega^2}\, d\omega, \tag{4.39}$$

$$-B(\lambda) = \lim_{\sigma\to 0} \int_{-\infty}^{\lambda} \frac{\log^-|F(\sigma+i\omega)|}{1+\omega^2}\, d\omega.$$

By hypothesis, $F_1(p)$ is H^n-stable:

$$\int_{-\infty}^{\infty} |F_1(\sigma+i\omega)|^n\, d\omega < C \quad \text{for} \quad \sigma > 0$$

where C is a fixed, positive constant. Hence, uniformly for $\sigma > 0$,

$$\int_{-\infty}^{\lambda} |F_1(\sigma+i\omega)|^n\, d\omega$$

is an absolutely continuous distribution function. Because, for $a \geqslant 0$,

$$0 \leqslant \log^+ a \leqslant a,$$

we have

$$0 \leqslant \log^+|F_1(\sigma+i\omega)|^n \leqslant |F_1(\sigma+i\omega)|^n.$$

Since $1+\omega^2 \geqslant 1$, we have

$$0 \leqslant \frac{n\log^+|F_1(\sigma+i\omega)|}{1+\omega^2} \leqslant \log|F_1(\sigma+i\omega)|^n.$$

Hence, since $n > 0$,

$$0 \leqslant \frac{\log^+ |F_1(\sigma+i\omega)|}{1+\omega^2} \leqslant \frac{1}{n} |F_1(\sigma+i\omega)|^n.$$

Therefore, uniformly for $\sigma > 0$,

$$\int_{-\infty}^{\lambda} \frac{\log^+ |F_1(\sigma+i\omega)|}{1+\omega^2} \, d\omega$$

is an absolutely continuous distribution function, and hence

$$\lim_{\sigma \to 0} \int_{-\infty}^{\lambda} \frac{\log^+ |F_1(\sigma+i\omega)|}{1+\omega^2} \, d\omega$$

is an absolutely continuous distribution function. By equation (4.39) this function is $\Gamma(\lambda)$, and so $\Gamma(\lambda)$ is absolutely continuous. In the representation of $\Gamma(\lambda)$, we therefore have $\gamma_j = 0$, which gives

$$\Gamma(\lambda) = \int_{-\infty}^{\lambda} \gamma(\omega) \, d\omega = \int_{-\infty}^{\lambda} \frac{\log^+ |F(i\omega)|}{1+\omega^2} \, d\omega.$$

We finally have

$$\Psi(\lambda) = \lim_{\sigma \to 0} \int_{-\infty}^{\lambda} \frac{\log |F_1(\sigma+i\omega)|}{1+\omega^2} \, d\omega$$

$$= \int_{-\infty}^{\lambda} [\gamma(\omega) - \beta(\omega) - \sum_k \beta_k \, \delta(\omega - \lambda_k)] \, d\omega$$

where the density function $\gamma(\omega)$ is

$$\gamma(\omega) = \frac{\log^+ |F_1(i\omega)|}{1+\omega^2},$$

the density function $\beta(\omega)$ is

$$\beta(\omega) = -\frac{\log^- |F_1(i\omega)|}{1+\omega^2},$$

and the β_k are arbitrary, where

$$\beta_k \geqslant 0, \quad \sum_k \beta_k < \infty.$$

More concisely,

$$\Psi(\lambda) = \int_{-\infty}^{\lambda} \frac{\log |F_1(i\omega)|}{1+\omega^2} \, d\omega - \sum_k \beta_k \, u(\lambda - \lambda_k)$$

and

$$d\Psi(\lambda) = \frac{\log |F_1(i\lambda)|}{1+\lambda^2} \, d\lambda - \sum_k \beta_k \, \delta(\lambda - \lambda_k).$$

Equation (4.34) becomes

$$\log F_1(p) = \frac{1}{\pi} \int_{-\infty}^{\infty} \frac{1 - i\lambda p}{p - i\lambda} \frac{\log |F_1(i\lambda)|}{1+\lambda^2} \, d\lambda$$

$$- \frac{1}{\pi} \int_{-\infty}^{\infty} \frac{1 - i\lambda p}{p - i\lambda} \sum_k \beta_k \, \delta(\lambda - \lambda_k) + ic.$$

We allow the finite jump $\beta_{-\infty}$ at $\lambda_{-\infty}=-\infty$ and the finite jump β_∞ at $\lambda_\infty=\infty$. If we define α to be $\alpha=(\beta_{-\infty}+\beta_\infty)/\pi$, then $0\leqslant\alpha<\infty$. By the convolution property of the impulse function, we have

$$-\frac{1}{\pi}\int_{-\infty}^{\infty}\frac{1-i\lambda p}{p-i\lambda}\sum_k\beta_k\,\delta(\lambda-\lambda_k)=-\frac{1}{\pi}\sum_k\frac{1-i\lambda_k p}{p-i\lambda_k}\beta_k,$$

which, by requiring all the λ_k to be finite, is equal to

$$-\frac{1}{\pi}\sum_k\frac{1-\lambda_k p}{p-i\lambda_k}\beta_k-p\alpha.$$

Hence,

$$\begin{aligned}F_1(p)&=\exp[\log F_1(p)]\\&=\exp\left[\frac{1}{\pi}\int_{-\infty}^{\infty}\frac{1-i\lambda p}{p-i\lambda}\frac{\log|F_1(i\lambda)|}{1+\lambda^2}d\lambda\right]\\&\quad\times\exp\left[-\frac{1}{\pi}\sum_k\frac{1-i\lambda_k p}{p-i\lambda_k}\right]\exp(-p\alpha)\exp(ic)\\&=F_0(p)A_3(p)A_2(p)A_0.\end{aligned}$$

Q.E.D.

Theorem 4.31. Let $F_1(p)$ and $F_0(p)$ be given as in Theorem 4.30, and let $\alpha>0$ and/or some $\beta_k>0$. Then, in the right half-plane,

$$|F_1(p)|<|F_0(p)|;$$

and, almost everywhere on the $i\omega$ axis,

$$|F_1(i\omega)|=|F_0(i\omega)|=M(i\omega).$$

Proof. The first conclusion follows from equations (4.31), (4.32), and (4.33). Letting $\sigma\to 0$ in equations (4.31) and (4.32), we have

$$|A_2(i\omega)|=1,\quad |A_3(i\omega)|=1$$

almost everywhere on the $i\omega$ axis. (Note: the points λ_k are countable and so represent a set of Lebesgue measure zero.) Hence, for almost all ω,

$$|F_1(i\omega)|=|F_0(i\omega)|.$$

The attenuation is

$$\begin{aligned}\log|F_0(i\omega)|&=\lim_{\sigma\to 0}\log|F_0(i\omega+\sigma)|\quad(\sigma>0)\\&=\lim_{\sigma\to 0}\int_{-\infty}^{\infty}[\operatorname{Re}K(p;\lambda)]\log M(i\lambda)\,d\lambda\\&=\int_{-\infty}^{\infty}\left[\lim_{\sigma\to 0}\operatorname{Re}K(p;\lambda)\right]\log M(i\lambda)\,d\lambda\\&=\int_{-\infty}^{\infty}\delta(\omega-\lambda)\log M(i\lambda)\,d\lambda=\log M(i\omega)\end{aligned}$$

where we have used Lemma 4.13. Hence

$$|F_0(i\omega)| = M(i\omega)$$

for almost all ω. Q.E.D.

In the above proof, we have also proved the following two theorems:

Theorem 4.32. Let $A_2(p)$ be given as in Theorem 4.30(e), and let $\alpha > 0$. Then, in the right half-plane,

(a) $A_2(p)$ is analytic,
(b) $A_2(p)$ has no zeros,
(c) $|A_2(p)| < 1$;

and, almost everywhere on the $i\omega$ axis,

$$|A_2(i\omega)| = 1.$$

Theorem 4.33. Let $A_3(p)$ be given as in Theorem 4.30(f), and let some $\beta_k > 0$. Then, in the right half-plane,

(a) $A_3(p)$ is analytic,
(b) $A_3(p)$ has no zeros,
(c) $|A_3(p)| < 1$;

and, almost everywhere on the $i\omega$ axis,

(d) $|A_3(i\omega)| = 1$.

Example 4.31. An example of $A_3(p)$ is $e^{(-1/p)}$ which is obtained by letting all $\beta_k = 0$, except $\beta_0 = \pi$ for $\lambda_0 = 0$.

4.4. The canonical representation

Theorems 4.20 and 4.30 may be combined to give:

Theorem 4.40 (the canonical representation of an H^n-stable transfer function). A non-null transfer function $F(p)$

(a) is analytic in the right half-plane,
(b) is H^n-stable,
(c) has zeros $\{p_k\}$, $\operatorname{Re} p_k > 0$, in the right half-plane

if and only if $F(p)$ has the representation

$$F(p) = A_0 A_1(p) A_2(p) A_3(p) F_0(p)$$

where A_0, $A_1(p)$, $A_2(p)$, and $A_3(p)$, $F_0(p)$ are given as in Theorems 4.20 and 4.30. This representation of $F(p)$ is unique.

This canonical representation was established by V. I. Krylov (1939) in the Russian language; independently by K. Karhunen (1949) in the German language, and again independently by C. Wilcox (1959) in the English language. Their proofs make use of the conformal transformation of the half-plane into the unit circle, so that theorems about functions analytic within the unit circle become applicable.

CHAPTER V

ALL-PASS SYSTEMS

5.1. All-pass systems

The canonical representation, Theorem 4.40, motivates the following definitions of the various types of all-pass systems.

Definition 5.10. Let A_0 be a constant of modulus $|A_0| = 1$. Then A_0 is called a *Type 0*, or *no-delay*, or *trivial*, *all-pass* transfer function.

Definition 5.11. Let

$$A_1(p) = \prod_k \frac{p_k - p}{\bar{p}_k + p} \frac{|p_k - 1|}{p_k - 1} \frac{|p_k + 1|}{p_k + 1},$$

with the $\{p_k\}$ satisfying $\operatorname{Re} p_k > 0$ and

$$\sum_k \frac{\operatorname{Re} p_k}{1 + |p_k|^2} < \infty.$$

Then if the set $\{p_k\}$ is not empty, $A_1(p)$ is called a *Type 1*, or *lumped-delay*, *all-pass* transfer function.

Definition 5.12. Let

$$A_2(p) = \mathrm{e}^{-\alpha p}, \quad \alpha \geqslant 0.$$

Then, if $\alpha > 0$, $A_2(p)$ is called a *Type 2*, or *pure-delay*, *all-pass* transfer function. The quantity α is called the *time-delay*.

Note: The pure-delay system has been defined previously (Definition 3.42).

Example 5.10. We have the approximations

$$1 - \frac{\alpha p}{2} \approx \exp\left(-\frac{\alpha p}{2}\right), \quad \frac{1}{1 + \alpha p/2} \approx \exp\left(-\frac{\alpha p}{2}\right)$$

and hence

$$\frac{1 - \alpha p/2}{1 + \alpha p/2} \approx \mathrm{e}^{-\alpha p}.$$

The fraction is a rational approximation of a pure-delay transfer function and is of the form $A_0 A_1(p)$ where $|A_0| = 1$ and $A_1(p)$ is a lumped-delay transfer function with $p_1 = 2/\alpha$. A better approximation is

$$\left(\frac{2n/\alpha - p}{2n/\alpha + p}\right)^n = \left(\frac{1 - \alpha p/2n}{1 + \alpha p/2n}\right)^n \approx \mathrm{e}^{-\alpha p} \quad (n > 1).$$

Still other approximations of $e^{-\alpha p}$ by $A_0 A_1(p)$ may be found by allowing the p_k to be complex in conjugate pairs.

Definition 5.13. Let

$$A_3(p) = \exp\left[-\frac{1}{\pi}\sum_k \frac{1-i\lambda_k p}{p-i\lambda_k}\beta_k\right]$$

with λ_k real, $\beta_k \geqslant 0$, and $\sum_k \beta_k < \infty$. Then, if some $\beta_k > 0$, $A_3(p)$ is called a *Type 3*, or *impure-delay, all-pass* transfer function.

Example 5.11. If all $\beta_k = 0$ except $\beta_0 = \pi$ for $\lambda_0 = 0$, we obtain the impure-delay transfer function $e^{-1/p}$.

Definition 5.14. Let

$$A(p) = A_0 A_1(p) A_2(p) A_3(p),$$

where the factor $A_j(p)$ is a type j all-pass system, and where any or several of these factors may be absent. Then $A(p)$ is called an *all-pass* transfer function. If all the factors except A_0 are absent, then $A(p)$ is called *trivial*; otherwise $A(p)$ is called *non-trivial.*

5.2. Admissible L^n-stable gain and the central minimum-delay transfer function

The canonical representation, Theorem 4.40, motivates the following definitions:

Definition 5.20. If $M(i\lambda) \geqslant 0$, $M(i\lambda)$ is L^n-stable, and

$$\int_{-\infty}^{\infty} \frac{\log M(i\lambda)}{1+\lambda^2}\, d\lambda > -\infty,$$

then $M(i\lambda)$ is called an *admissible, L^n-stable gain.*

Definition 5.21. If $M(i\lambda)$ is an admissible, L^n-stable gain, then

$$F_0(p) = \exp\left[\frac{1}{\pi}\int_{-\infty}^{\infty} \frac{1-i\lambda p}{p-i\lambda}\,\frac{\log M(i\lambda)}{1+\lambda^2}\, d\lambda\right]$$

is called the *central minimum-delay transfer function* for $M(i\lambda)$. If $f_0(t) \leftrightarrow F_0(p)$, then $f_0(t)$ is called the *central minimum-delay time-function* for $M(i\lambda)$.

If $K(p;\lambda)$ denotes the kernel function of Definition 4.12, then the central minimum-delay transfer function may be written more compactly as

$$F_0(p) = \exp\left[\int_{-\infty}^{\infty} K(p;\lambda) \log M(i\lambda)\, d\lambda\right].$$

The canonical representation now becomes

Theorem 5.20 (*the canonical representation*). A non-null transfer function $F(p)$ is analytic in the right half-plane and H^n-stable if and only if

$$F(p) = A(p)F_0(p),$$

where $A(p)$ is an all-pass transfer function and $F_0(p)$ is the central minimum-delay transfer function for the admissible, L^n-stable gain $M(i\lambda)$. This representation of $F(p)$ is unique.

Theorems 4.21 and 4.31 become:

Theorem 5.21. Let

$$F(p) = A(p)F_0(p)$$

be the canonical representation of the transfer function $F(p)$ that is analytic in the right half-plane and H^n-stable. Then, almost everywhere on the $i\omega$-axis, we have

$$|F(i\omega)| = |F_0(i\omega)| = M(i\omega).$$

If $A(p)$ is trivial, then $|F(p)| = |F_0(p)|$ in the right half-plane. If $A(p)$ is non-trivial, then $|F(p)| < |F_0(p)|$ in the right half-plane.

Theorems 4.22, 4.32, and 4.33 become:

Theorem 5.22. If $A(p)$ is a trivial all-pass transfer function, then $A(p) = A_0$ where $|A_0| = 1$. If $A(p)$ is a non-trivial all-pass transfer function, then, in the right half-plane,

(a) $A(p)$ is analytic,
(b) $A(p)$ has zeros $\{p_k\}$ (where the set $\{p_k\}$ may be empty),
(c) $|A(p)| < 1$,

and, almost everywhere on the $i\omega$-axis,

$$|A(i\omega)| = 1.$$

5.3. Wavelets and all-pass systems

Theorem 3.10 states that $W(p)$ is the transfer function of a wavelet if and only if $W(p)$ is analytic in the right half-plane and H^2-stable. Letting $n = 2$ in the canonical representation, Theorem 5.20, we therefore have

Theorem 5.30 (*the canonical representation of the transfer function of a non-null wavelet*). $W(p)$ is the transfer function of a non-null wavelet if and only if

$$W(p) = A(p)W_0(p),$$

where $A(p)$ is an all-pass transfer function and $W_0(p)$ is the central minimum-delay transfer function for the admissible, L^2-stable gain $M(i\omega)$. This representation of $W(p)$ is unique.

Theorem 5.31. Let a wavelet $w(t)$ be the input to an all-pass system $D(p)$. Then the resulting output $v(t)$ is a wavelet with the same gain as $w(t)$.

Proof. Let $w(t) \leftrightarrow W(p)$. Then, by Theorem 5.30, $W(p)$ has the canonical representation

$$W(p) = A(p)\, W_0(p).$$

Since

$$w(t) \to \boxed{\text{System } D(p)} \to v(t),$$

we have

$$v(t) \leftrightarrow V(p) = D(p)\, W(p).$$

Hence

$$V(p) = D(p)\, A(p)\, W_0(p),$$

which is the canonical representation of $V(p)$ since $D(p)\,A(p)$ is an all-pass transfer function. Hence $v(t)$ is a wavelet with gain

$$|V(i\omega)| = |W(i\omega)| = |W_0(i\omega)|. \qquad \text{Q.E.D.}$$

5.4. Realizability and unit gain property of all-pass systems

Theorem 5.40. A system is realizable and has gain equal to one for almost all frequencies ω if and only if the system is an all-pass system.

Proof ("if" statement). By hypothesis $A(p)$ is an all-pass system. By Theorem *5.22*, $|A(i\omega)| = 1$ for almost all ω. By Definition 5.14, we have

$$A(p) = A_0\, A_1(p)\, A_2(p)\, A_3(p),$$

where $A_j(p)$ is a Type j all-pass system ($j = 0, 1, 2, 3$). Let

$$a_j(t) \leftrightarrow A_j(p),$$

i.e. let $a_j(t)$ be the impulse response of the Type j all-pass system. In particular, we have

$$a_0(t) = A_0\, \delta(t), \quad |A_0| = 1,$$

which is realizable. $A_1(p)$ is the product of Type 0 all-pass systems and factors of the form

$$\frac{\gamma - p}{\bar{\gamma} + p}$$

where $\operatorname{Re} p > 0$, $\gamma = \zeta + i\eta$, $\zeta > 0$. This factor may be written

$$\frac{\gamma - p}{\bar{\gamma} + p} = -1 + \frac{2\zeta}{\bar{\gamma} + p},$$

which corresponds to the time-function

$$-\delta(t)+2\zeta u(t)\,e^{-\bar{\gamma}t},$$

which is realizable. We have

$$a_2(t) = \delta(t-\alpha) \leftrightarrow e^{-\alpha p},$$

where $\mathrm{Re}\,p>0$, $\alpha>0$. Because $\delta(t-\alpha) = 0$ for $t<0$, $a_2(t)$ is realizable. It may also be shown that $a_3(t)$ is realizable. Letting $a(t) \leftrightarrow A(p)$, we have

$$a(t) = a_0(t) * a_1(t) * a_2(t) * a_3(t),$$

and since the convolution of one-sided time-functions is one-sided, it follows that $a(t)$ is one-sided, i.e. realizable. Q.E.D. (1).

Proof ("only if" statement). By hypothesis, $D(p)$ has gain $|D(i\omega)| = 1$ and $d(t) \leftrightarrow D(p)$ where $d(t)$ is one-sided. Consider the central minimum-delay transfer function $W_0(p)$ for the admissible, L^2-stable gain $M(i\omega)$. If $w_0(t) \leftrightarrow W_0(p)$, it follows by Theorem 5.30 that $w_0(t)$ is a wavelet. Let $w_0(t)$ be the input to the system $D(p)$ and let $g(t)$ be the resulting output. Let $g(t) \leftrightarrow G(p)$. Then

$$G(p) = D(p)\,W_0(p),$$

which gives

$$|G(i\omega)| = |D(i\omega)|\,|W_0(i\omega)| = |W_0(i\omega)| = M(i\omega),$$

and hence $|G(i\omega)|$ is L^2-stable. Since

$$\int_{-\infty}^{\infty} |g(t)|^2\,dt = \frac{1}{2\pi}\int_{-\infty}^{\infty} |G(i\omega)|^2\,d\omega,$$

it follows that $g(t)$ is L^2-stable. Since, by hypothesis, $d(t)$ is one-sided, it follows that

$$g(t) = d(t) * w_0(t)$$

is one-sided. Hence $g(t)$ is a one-sided, L^2-stable time-function, i.e. $g(t)$ is a wavelet. Hence it follows that

$$G(p) = D(p)\,W_0(p)$$

is the canonical representation of $G(p)$, and so $D(p)$ is an all-pass system. Q.E.D.

5.5. Group-delay of an all-pass system

An all-pass system $A(p)$ by Theorem 5.22 has gain

$$|A(i\omega)| = 1$$

almost everywhere on the $i\omega$-axis. Hence the *gain-and-phase spectrum of a Type j all-pass system* may be written

$$A_j(i\omega) = \exp[iQ_j(i\omega)], \quad j = 0, 1, 2, 3.$$

Letting $p = i\omega$ in Definitions 5.10, 5.11, 5.12, and 5.13, we find, respectively, the phase-shifts:

$$Q_0 = c \ (c \text{ a real constant}),$$

$$Q_1(i\omega) = \sum_k \left[2\tan^{-1}\frac{\eta_k - \omega}{\zeta_k} - \tan^{-1}\frac{\eta_k}{\zeta_k - 1} - \tan^{-1}\frac{\eta_k}{\zeta_k + 1} \right]$$

$$(\text{where } p_k = \zeta_k + i\eta_k,\ \zeta_k > 0),$$

$$Q_2(i\omega) = -\alpha\omega \quad (\alpha > 0),$$

$$Q_3(i\omega) = \frac{1}{\pi}\sum_k \frac{1+\lambda_k\omega}{\omega - \lambda_k}\beta_k.$$

The gain-and-phase spectrum of an all-pass system then may be written

$$A(i\omega) = \exp[iQ(i\omega)] = \exp[i\sum_j Q_j(i\omega)],$$

where the Q_j are given above.

Theorem 5.50. The group-delay of a trivial all-pass system is zero over the entire frequency range.

Proof. A trivial all-pass system has phase-shift c, where c is a real constant. Hence its group-delay is

$$\tau_g = -\frac{dc}{d\omega} = 0, \quad -\infty < \omega < \infty. \qquad \text{Q.E.D.}$$

Theorem 5.51. The group-delay of an all-pass system is positive over the entire frequency range if and only if the all-pass system is non-trivial.

Proof ("if" statement). By hypothesis, the all-pass system is non-trivial. Hence its group delay is

$$\tau_g = -\frac{dQ(i\omega)}{d\omega} = \sum_j \left[-\frac{dQ_j(i\omega)}{d\omega} \right],$$

where at least one of the phase-shifts $Q_1(i\omega)$, $Q_2(i\omega)$, $Q_3(i\omega)$ is not absent. We have

$$-\frac{dQ_1(i\omega)}{d\omega} = \sum_k \frac{2\zeta_k}{\zeta_k^2 + (\eta_k - \omega)^2} > 0,$$

$$-\frac{dQ_2(i\omega)}{d\omega} = \alpha > 0,$$

$$-\frac{dQ_3(i\omega)}{d\omega} = \frac{1}{\pi}\sum_k \frac{\lambda_k^2 + 1}{(\omega - \lambda_k)^2}\beta_k > 0. \qquad \text{Q.E.D.}$$

Proof ("only if" statement). Clear. Q.E.D.

5.6. Delay of energy by an all-pass system

Theorem 5.31 tells us that an all-pass system passes an input wavelet $w(t)$ into an output wavelet $v(t)$ with the same gain: $|W(i\omega)| = |V(i\omega)|$. Consequently,

$$\frac{1}{2\pi}\int_{-\infty}^{\infty} |W(i\omega)|^2\, d\omega = \frac{1}{2\pi}\int_{-\infty}^{\infty} |V(i\omega)|^2\, d\omega,$$

and so using Bessel's equality (Theorem 2.13), we have

$$\int_0^{\infty} |w(t)|^2\, dt = \int_0^{\infty} |v(t)|^2\, dt,$$

which says that input and output wavelets have the *same total energy*. The following theorem, however, tells us that the *partial energy* of the output wavelet *is delayed* with respect to the partial energy of the input.

Theorem 5.60 (*the partial energy theorem*). Let a wavelet $w(t)$ be the input to an all-pass system $A(p)$ and let the wavelet $v(t)$ be the resulting output. If the all-pass system is trivial, the partial energy of input is equal to the partial energy of output, i.e.

$$\int_0^{\tau} |w(t)|^2\, dt = \int_0^{\tau} |v(t)|^2\, dt.$$

If the all-pass system is non-trivial, the partial energy of input is greater than or equal to the partial energy of output, i.e.

$$\int_0^{\tau} |w(t)|^2\, dt \geqslant \int_0^{\tau} |v(t)|^2\, dt,$$

and the strict inequality (i.e. *greater than*) holds for some $\tau > 0$.

Proof. (1) If the all-pass system is trivial, then

$$A_0\, w(t) = v(t), \quad |A_0| = 1,$$

from which the desired result immediately follows.

(2) If the all-pass system is Type 2, then

$$v(t) = w(t-\alpha), \quad \alpha > 0,$$

and

$$\int_0^{\tau} |w(t)|^2\, dt = \int_{\alpha}^{\tau+\alpha} |w(t-\alpha)|^2\, dt = \int_0^{\tau+\alpha} |w(t-\alpha)|^2\, dt,$$

since $w(t-\alpha) = 0$ for $t < \alpha$. Hence

$$\int_0^{\tau} |w(t)|^2\, dt > \int_0^{\tau} |w(t-\alpha)|^2\, dt = \int_0^{\tau} |v(t)|^2\, dt$$

for some $\tau > 0$.

(3) If the system is Type 1, then it is the product of a trivial all-pass system A_0, $|A_0| = 1$, and factors of the type

$$\frac{\gamma - p}{\bar{\gamma} + p}, \quad \gamma = \zeta + i\eta, \quad \zeta > 0, \quad \operatorname{Re} p > 0,$$

which (see Section **5.4**) has impulse response

$$-\delta(t) + 2\zeta u(t)\,\mathrm{e}^{-\bar{\gamma}t}.$$

Hence

$$v(t) = w(t) * [-\delta(t) + 2\zeta u(t)\,\mathrm{e}^{-\bar{\gamma}t}] = -w(t) + 2\zeta \int_0^t \mathrm{e}^{-\bar{\gamma}(t-s)}\, w(s)\, ds.$$

Letting $t = 0$, the following initial condition results:

$$v(0) = -w(0).$$

If $W(p)$ and $V(p)$ are the Laplace transforms of input and output, respectively, then

$$\frac{\gamma - p}{\bar{\gamma} + p} W(p) = V(p),$$

which gives

$$(\gamma - p)\, W(p) = (\bar{\gamma} + p)\, V(p).$$

Using our operational rules (see Section **2.5**), we have

$$\gamma w(t) - w'(t) = \bar{\gamma} v(t) + v'(t),$$

which gives

$$w'(t) + v'(t) = \gamma w(t) - \bar{\gamma} v(t).$$

The following identity is true, as may be readily verified by expansion (where $\gamma = \zeta + i\eta$):

$$2\zeta[|w|^2 - |v|^2]$$
$$\equiv (\bar{\gamma} + \gamma)\, w\bar{w} - (\bar{\gamma} + \gamma)\, v\bar{v} \equiv (w + v)\,\overline{(\gamma w - \bar{\gamma} v)} + \overline{(w + v)}\,(\gamma w - \bar{\gamma} v).$$

Because

$$w' + v' = \gamma w - \bar{\gamma} v,$$

we have

$$2\zeta[|w|^2 - |v|^2] = (w + v)\,\overline{(w' + v')} + \overline{(w + v)}\,(w' + v')$$
$$= \frac{d}{dt}[(w + v)\,\overline{(w + v)}] = \frac{d}{dt}[|w + v|^2].$$

Integrating, we have

$$2\zeta\int_0^\tau [|w(t)|^2-|v(t)|^2]\,dt = \int_0^\tau \frac{d}{dt}[|w(t)+v(t)|^2]\,dt$$
$$= |w(\tau)+v(\tau)|^2 - |w(0)+v(0)|^2$$
$$= |w(\tau)+v(\tau)|^2$$

because of the initial condition $v(0) = -w(0)$. Hence

$$\int_0^\tau |w(t)|^2\,dt = \int_0^\tau |v(t)|^2\,dt + \frac{1}{2\zeta}|w(\tau)+v(\tau)|^2 > \int_0^\tau |v(t)|^2\,dt$$

since $\zeta > 0$.

(4) If the all-pass system is Type 3, it may also be shown that the desired result is true. Q.E.D.

5.7. Inverse to an all-pass system

Theorem 5.70. The inverse system to an all-pass system has unit gain for almost all frequencies ω. The inverse system to a trivial all-pass system is realizable (and hence is an all-pass system). The inverse system to a non-trivial all-pass system is non-realizable (and hence is not an all-pass system).

Proof. Let $A(p)$ be an all-pass transfer function. Then $|A(i\omega)| = 1$ almost everywhere on the $i\omega$-axis, and

$$|A^{-1}(i\omega)| = \frac{1}{|A(i\omega)|} = 1$$

almost everywhere on the $i\omega$-axis. If $A(p) = A_0$, $|A_0| = 1$, then

$$A_0^{-1} = \frac{1}{A_0}, \quad |A_0^{-1}| = 1,$$

and so A_0^{-1} is also a trivial all-pass system. If $A(p)$ is non-trivial, then by Definition 5.14,

$$A(p) = A_0 A_1(p) A_2(p) A_3(p),$$

where at least one of the factors $A_1(p)$, $A_2(p)$, $A_3(p)$ is not absent. Then, clearly,

$$A^{-1}(p) = \frac{1}{A_0 A_1(p) A_2(p) A_3(p)}$$

violates Definition 5.14, and so $A^{-1}(p)$ cannot be an all-pass transfer function. Furthermore, $A^{-1}(p)$ cannot be the transfer function of a realizable system, for if $A^{-1}(p)$ were, then Theorem 5.40 would be contradicted. Q.E.D.

Example 5.70. The inverse system to the Type 0 all-pass system with transfer function $A_0, |A_0| = 1$, has transfer function $A_0^{-1} = 1/A_0$ and (realizable) impulse response

$$a_0^{-1}(t) = \frac{1}{A_0}\,\delta(t).$$

Example 5.71. The inverse system to the Type 1 all-pass system with the transfer function

$$A_1(p) = C_0\frac{\gamma - p}{\bar{\gamma} + p}, \quad C_0 = \frac{|\gamma - 1|}{\gamma - 1}\,\frac{|\gamma + 1|}{\gamma + 1},$$

where $\gamma = \zeta + i\eta$, $\zeta > 0$, Re $p > 0$, has transfer function

$$A_1^{-1}(p) = \frac{1}{C_0}\,\frac{\bar{\gamma} + p}{\bar{\gamma} - p} = \frac{1}{C_0}\left[-1 + \frac{2\zeta}{\gamma - p}\right], \quad \operatorname{Re} p < \alpha,$$

and (non-realizable) impulse response

$$a_1^{-1}(t) = \frac{1}{C_0}[-\delta(t) + 2\zeta u(-t)\,e^{\gamma t}].$$

Example 5.72. The inverse system to the Type 2 all-pass system with transfer function

$$A_2(p) = e^{-\alpha p}, \quad \alpha > 0$$

has transfer function

$$A_2^{-1}(p) = e^{\alpha p}, \quad \alpha > 0,$$

and (non-realizable) impulse response

$$a_2^{-1}(t) = \delta(t + \alpha)$$

which represents a pure advance.

CHAPTER VI

THE KARHUNEN THEOREM

6.1. The canonical wavelet representation

Because a realizable, dissipative system is a system whose impulse response is a wavelet, Theorem 5.30 becomes

Theorem 6.10 (*the canonical representation of a non-null, realizable, dissipative transfer function*). $W(p)$ is the transfer function of a non-null, realizable, dissipative system if and only if

$$W(p) = A(p) W_0(p)$$

where $A(p)$ is an all-pass transfer function and $W_0(p)$ is the central minimum-delay transfer function for the admissible, L^2-stable gain $M(i\omega)$. This representation of $W(p)$ is unique.

If $w_0(t) \leftrightarrow W_0(p)$, then $w_0(t)$ is called the *central minimum-delay wavelet* for the admissible, L^2-stable gain $M(i\omega)$. We have

Theorem 6.11 (*the canonical wavelet representation*). $w(t)$ is a non-null wavelet if and only if

$$w(t) = a(t) * w_0(t)$$

where $a(t)$ is an all-pass impulse response function and $w_0(t)$ is the central minimum-delay wavelet for the admissible, L^2-stable gain $M(i\omega)$. This representation of $w(t)$ is unique.

6.2. Auto-products and cross-products of wavelets

Definition 6.20. Let $f(t)$ be a time-function. Its *anti-function* is defined to be $\overline{f(-t)}$.

For example, if $w(t)$ is a wavelet, its anti-wavelet is $\overline{w(-t)}$. Clearly an *anti-wavelet* $\overline{w(-t)}$ vanishes for positive time t and is L^2-stable. If $b(t)$ is a weighting function, then the *anti-weighting function* is $\overline{b(-t)}$.

Definition 6.21. The *auto-product* $\phi(t) \equiv f(t) \otimes f(t)$ of a time function $f(t)$ is defined to be the convolution of the function $f(t)$ with its anti-function $\overline{f(-t)}$.

The auto-product may be written as

$$\begin{aligned}\phi(t) &= f(t) \otimes f(t) = f(t) * \overline{f(-t)} \\ &= \int_{-\infty}^{\infty} f(t-\tau)\overline{f(-\tau)}\,d\tau = \int_{-\infty}^{\infty} f(t+s)\overline{f(s)}\,ds.\end{aligned}$$

Theorem 6.20. Let $\phi(t)$ be the auto-product of $f(t)$. Then the anti-auto-product of $\phi(t)$ is equal to $\phi(t)$; that is,

$$\overline{\phi(-t)} = \phi(t).$$

Proof. Letting $\tau = t+s$, we have

$$\phi(t) = \int_{-\infty}^{\infty} f(t+s)\overline{f(s)}\,ds = \int_{-\infty}^{\infty} f(\tau)\overline{f(\tau-t)}\,d\tau$$

$$= \overline{\left(\int_{-\infty}^{\infty} f(-t+\tau)\overline{f(\tau)}\,d\tau\right)} = \overline{\phi(-t)}. \qquad \text{Q.E.D.}$$

We note that convolution of a wavelet $w(t)$ with itself is the one-sided time-function given by

$$w(t) * w(t) = \int_0^{\infty} w(t-\tau)\,w(\tau)\,d\tau = 0 \quad \text{for} \quad t<0,$$

whereas the auto-product of a wavelet $w(t)$ is the two-sided time-function

$$\phi(t) = w(t) \otimes w(t) = \int_{-\infty}^{0} w(t-\tau)\,\overline{w(-\tau)}\,d\tau = \int_0^{\infty} w(t+s)\,\overline{w(s)}\,ds$$

where $\phi(-t) = \overline{\phi(t)}$.

Theorem 6.21. Let $W(p)$ be the Laplace transform of the wavelet $w(t)$, and let $F(p)$ be the Laplace transform of the anti-wavelet $f(t) \equiv \overline{w(-t)}$. Then

$$F(p) = \overline{W(-\bar{p})}.$$

Proof. For $\sigma = \operatorname{Re} p \leqslant 0$, we have

$$F(p) = \int_{-\infty}^{0} f(t)\,\mathrm{e}^{-pt}\,dt = \int_{-\infty}^{0} \overline{w(-t)}\,\mathrm{e}^{-pt}\,dt$$

$$= \int_0^{\infty} \overline{w(\tau)}\,\mathrm{e}^{p\tau}\,d\tau = \overline{\left(\int_0^{\infty} w(\tau)\,\overline{\mathrm{e}^{p\tau}}\,d\tau\right)}$$

$$= \overline{\left(\int_0^{\infty} w(\tau)\,\mathrm{e}^{\bar{p}\tau}\,d\tau\right)} = \overline{\left(\int_0^{\infty} w(\tau)\,\mathrm{e}^{-(-\bar{p})\tau}\,d\tau\right)} = \overline{W(-\bar{p})},$$

$$\operatorname{Re}(-\bar{p}) = -\sigma = -\operatorname{Re} p \geqslant 0,$$

where W is given by

$$W(p') = \int_0^{\infty} w(\tau)\,\mathrm{e}^{-p'\tau}\,d\tau, \quad \operatorname{Re} p' > 0, \quad p' = -\bar{p}. \qquad \text{Q.E.D.}$$

Corollary 6.20. Two wavelets have the same gain if and only if they have the same auto-product.

Proof. Let $w_1(t)$ and $w_2(t)$ each have the gain $M(i\omega)$. Then, for $i = 1, 2$,

$$\phi_{ii}(s) = w_i(t) * \overline{w_i(-t)} \leftrightarrow W_i(i\omega)\,\overline{W_i(i\omega)} = M^2(i\omega). \qquad \text{Q.E.D.}$$

Definition 6.22. The *cross-product* $\phi_{12}(t) \equiv f_1(t) \otimes f_2(t)$ of the time-function $f_1(t)$ with the time-function $f_2(t)$ is defined to be the convolution of the function $f_1(t)$ with the anti-function of $f_2(t)$; that is,

$$\phi_{12}(t) = f_1(t) \otimes f_2(t) = f_1(t) * \overline{f_2(-t)}.$$

Note: The cross-product $\phi_{12}(t)$ may also be denoted by $\phi_{f_1 f_2}(t)$. We see that the auto-product of $f(t)$ is the cross-product $\phi_{ff}(t)$ of $f(t)$ with itself. The auto-product of $f(t)$ and $g(t)$ may be expressed as

$$\phi_{fg}(t) = f(t) \otimes g(t) = f(t) * \overline{g(-t)}$$

$$= \int_{-\infty}^{\infty} f(t-\tau)\,\overline{g(-\tau)}\,d\tau = \int_{-\infty}^{\infty} f(t+s)\,\overline{g(s)}\,ds.$$

Theorem 6.22. Let $\phi_{fg}(t)$ be the cross-product of $f(t)$ with $g(t)$. Then the anti-cross-product of $\phi_{gf}(t)$ is equal to $\phi_{fg}(t)$; that is,

$$\overline{\phi_{gf}(-t)} = \phi_{fg}(t).$$

Proof. Letting $\tau = t+s$, we have

$$\phi_{fg}(t) = \int_{-\infty}^{\infty} f(t+s)\,\overline{g(s)}\,ds = \int_{-\infty}^{\infty} f(\tau)\,\overline{g(\tau-t)}\,d\tau$$

$$= \overline{\left(\int_{-\infty}^{\infty} g(-t+\tau)\,\overline{f(\tau)}\,d\tau\right)} = \overline{\phi_{gf}(-t)}. \qquad \text{Q.E.D.}$$

6.3. The Karhunen theorem

An important theorem of Kari Karhunen (1949), expressed in terms of wavelets, is

Theorem 6.30 (the Karhunen theorem). Let $w(t)$ be a non-null wavelet and let $v(t)$ be either a null or non-null wavelet. Let their cross-product $w(t) \otimes v(t)$ be a one-sided time-function, i.e.

$$\int_{0}^{\infty} w(t+s)\,\overline{v(s)}\,ds = 0 \quad \text{for} \quad t < 0.$$

Then $v(t) \equiv 0$ if and only if

$$w(t) = A_0\, w_0(t)$$

where $|A_0| = 1$ and $w_0(t)$ is the central minimum-delay wavelet with the same gain as $w(t)$.

Proof. (1) Let us assume that $v(t)$ is non-null. By hypothesis $w(t)$ and $v(t)$ are wavelets, and hence $w(t)$ and $\overline{v(-t)}$ are L^2-stable. By Theorem 6.21, $\overline{V(i\omega)}$ is the Fourier transform of $\overline{v(-t)}$. It follows from the Convolution Theorem 2.16 that

$$\phi_{wv}(t) \equiv w(t) \otimes v(t) = w(t) * \overline{v(-t)} \tag{6.30}$$

is the L^1-Fourier transform of

$$\Phi_{wv}(i\omega) = W(i\omega)\overline{V(i\omega)}. \tag{6.31}$$

Since $\Phi_{wv}(i\omega)$ is L^1-stable (by Theorem 2.10) and its L^1-Fourier transform $\phi_{wv}(t)$ is one-sided (by hypothesis), it follows from Theorem 2.26 that

$$\Phi_{wv}(p) = \int_0^\infty \phi_{wv}(t)\, \mathrm{e}^{-pt}\, dt$$

is analytic in the right half-plane and H^1-stable. Therefore, by Theorem 5.20, $\Phi_{wv}(p)$ has the canonical representation

$$\Phi_{wv}(p) = A_{wv}(p)\,\Phi_{wv}^{(0)}(p) \tag{6.32}$$

where $A_{wv}(p)$ is an all-pass transfer function and $\Phi_{wv}^{(0)}(p)$ is the central minimum-delay transfer function

$$\Phi_{wv}^{(0)}(p) = \exp\left[\int_{-\infty}^{\infty} K(p;\lambda) \log|\Phi_{wv}(i\lambda)|\, d\lambda\right]. \tag{6.33}$$

(2) By assumption, $v(t)$ is a non-null wavelet. Therefore, by Theorem 5.30, its transfer function $V(p)$ has the canonical representation

$$V(p) = A_v(p)\, V_0(p) \tag{6.34}$$

where $A_v(p)$ is an all-pass transfer function and $V_0(p)$ is the central minimum-delay transfer function

$$V_0(p) = \exp\left[\int_{-\infty}^{\infty} K(p;\lambda) \log|V(i\lambda)|\, d\lambda\right]. \tag{6.35}$$

By hypothesis, $w(t)$ is a non-null wavelet, and so its transfer function has the canonical representation

$$W(p) = A_w(p)\, W_0(p) \tag{6.36}$$

where $A_w(p)$ is an all-pass transfer function and $W_0(p)$ is the central minimum-delay transfer function

$$W_0(p) = \exp\left[\int_{-\infty}^{\infty} K(p;\lambda) \log|W(i\lambda)|\, d\lambda\right]. \tag{6.37}$$

(3) We now define $\Phi(p)$ to be

$$\Phi(p) = \frac{V(p)\,\Phi_{wv}(p)}{W(p)}.$$

Since, using equation (6.31),

$$\Phi(i\omega) = \frac{V(i\omega)\,\Phi_{wv}(i\omega)}{W(i\omega)} = V(i\omega)\,\overline{V(i\omega)} = |V(i\omega)|^2 \geqslant 0, \qquad (6.38)$$

the gain of $\Phi(p)$ is

$$|\Phi(i\omega)| = \Phi(i\omega) = |V(i\omega)|^2.$$

Since, by assumption, $v(t)$ is a non-null wavelet, its gain $|V(i\omega)|$ is an admissible, L^2-stable gain. Hence the gain $\Phi(i\omega)$ is an admissible, L^1-stable gain. The central minimum-delay transfer function for $\Phi(i\omega)$ is then

$$\Phi_0(p) = \exp\left[\int_{-\infty}^{\infty} K(p;\lambda)\log\Phi(i\lambda)\,d\lambda\right]. \qquad (6.39)$$

(4) Using the canonical representations given by equations (6.32), (6.34), and (6.36), we have

$$\Phi(p) = \frac{A_v(p)\,A_{wv}(p)}{A_w(p)}\,\frac{V_0(p)\,\Phi_{wv}^{(0)}(p)}{W_0(p)}.$$

Using equations (6.33), (6.35), and (6.37) for the central minimum-delay transfer functions, we have

$$\frac{V_0(p)\,\Phi_{wv}^{(0)}(p)}{W_0(p)}$$

$$= \exp\left[\int_{-\infty}^{\infty} K(p;\lambda)\,[\log|V(i\lambda)| + \log|\Phi_{wv}(i\lambda)| - \log|W(i\lambda)|]\,d\lambda\right]$$

$$= \exp\left[\int_{-\infty}^{\infty} K(p;\lambda)\log\left|\frac{V(i\lambda)\,\Phi_{wv}(i\lambda)}{W(i\lambda)}\right|d\lambda\right]$$

which, by equations (6.38) and (6.39),

$$= \exp\left[\int_{-\infty}^{\infty} K(p;\lambda)\log|\Phi(i\lambda)|\,d\lambda\right] = \Phi_0(p).$$

(5) We now have

$$\Phi(p) = \frac{A_v(p)\,A_{wv}(p)}{A_w(p)}\,\Phi_0(p)$$

where $\Phi_0(p)$ is the central minimum-delay transfer function with the same gain as $\Phi(p)$. This representation for $\Phi(p)$ could be canonical if and only if

$$A_v(p)\,A_{wv}(p)/A_w(p)$$

is an all-pass transfer function, i.e. if and only if $A_w(p) = A_0$, where $|A_0| = 1$. By the Canonical Representation Theorem 5.20, $\Phi(p)$ has a canonical representation if and only if $\Phi(p)$ is analytic in the right half-plane and H^1-stable. By Theorems 2.26 and 2.27, $\Phi(p)$ is analytic in the right half-plane and H^1-stable if and only if

$$\Phi(p) = \int_0^\infty \phi(t)\, e^{-pt}\, dt$$

where $\phi(t)$ is a one-sided, bounded time function. Because

$$\Phi(i\omega) = V(i\omega)\, \overline{V(i\omega)}$$

it follows that

$$\phi(t) = \int_0^\infty v(t+s)\, \overline{v(s)}\, ds$$

and hence $\phi(t) = \overline{\phi(-t)}$. Thus $\phi(t)$ is one-sided if and only if $\phi(t)$ vanishes identically. Therefore $\phi(t)$ is null, and consequently our assumption that $v(t)$ is non-null is contradicted, if and only if $A_w(p) = A_0$ where $|A_0| = 1$, i.e. if and only if

$$W(p) = A_0 W_0(p), \quad |A_0| = 1. \qquad \text{Q.E.D.}$$

CHAPTER VII

THE MINIMUM-DELAY WAVELET THEOREM

7.1. Minimum-delay wavelets

The frequency-dependent characteristics of a system $F(p)$ that concern us are its attenuation and group-delay. A trivial all-pass system A_0 (i.e. a constant of modulus one) has zero attenuation and zero group-delay for all frequencies. Thus the systems $F(p)$ and $A_0 F(p)$ have exactly the same attenuation and the same group-delay, and so for our purposes are equivalent.

Definition 7.10. A wavelet (or realizable, dissipative system) $w(t)$ is called *minimum-delay* if $w(t) = A_0 w_0(t)$ where $|A_0| = 1$ and $w_0(t)$ is the central minimum-delay wavelet with the same gain as $w(t)$.

Clearly, a minimum-delay wavelet $w(t) \leftrightarrow W(p)$ is determined (except for a trivial all-pass system A_0) by its gain $|W(i\omega)|$ (or by its attenuation $\log|W(i\omega)|$); i.e. if $w(t)$ is minimum-delay, then

$$w(t) \leftrightarrow W(p) = A_0 \exp\left[\frac{1}{\pi}\int_{-\infty}^{\infty} \frac{1-i\lambda p}{p-i\lambda} \frac{\log|W(i\lambda)|}{1+\lambda^2} d\lambda\right].$$

7.2. Inverse wavelets

We now wish to continue the discussion of inverse systems given in Section **3.5**. A wavelet (or realizable, dissipative system) $w(t) \leftrightarrow W(i\omega)$ has an *inverse* (or is *invertible*) if and only if

$$W^{-1}(i\omega) = \frac{1}{W(i\omega)}$$

exists for almost all ω. A wavelet may be either null or non-null. If the wavelet is null, then $W(i\omega) = 0$ (for almost all ω), and no inverse exists. If the wavelet is non-null, then the canonical representation (Theorems 6.10 and 6.11) applies. Hence $|W(i\omega)|$ is an admissible, L^2-stable gain (Definition 5.20), and

$$\int_{-\infty}^{\infty} \frac{\log|W(i\lambda)|}{1+\lambda^2} d\lambda > -\infty.$$

The convergence of this integral requires that $|W(i\omega)| > 0$ (and hence $W(i\omega) \neq 0$) for almost all ω. Thus we have

Theorem 7.20. A wavelet is invertible if and only if it is non-null.

Examples 3.50, 3.52, and 3.53 illustrate special cases of the type of wavelet $w(t)$ whose transfer function $W(p)$ is a *rational function*

$$W(p) = \frac{N(p)}{D(p)}, \quad \operatorname{Re} p \geqslant 0,$$

where

$$\begin{aligned} N(p) &= b_n p^n + b_{n-1} p^{n-1} + \ldots + b_0 \\ &= b_n (p-\beta_1)(p-\beta_2) \ldots (p-\beta_n) \end{aligned}$$

with $b_n \neq 0$ and $\operatorname{Re}\beta_j \neq 0$ for $j = 1, 2, \ldots, n$; and where

$$\begin{aligned} D(p) &= a_m p^m + a_{m-1} p^{m-1} + \ldots + a_0 \\ &= a_m (p-\alpha_1)(p-\alpha_2) \ldots (p-\alpha_m) \end{aligned}$$

with $m > n$, $a_m \neq 0$, and $\operatorname{Re}\alpha_i < 0$ for $i = 1, 2, \ldots, m$.

We may assume that no α_i equals any β_j, for clearly any common factors could be cancelled out in the expression for $W(p)$. If $\alpha_1, \alpha_2, \ldots, \alpha_m$ are distinct, the transfer function has the partial fraction expansion

$$W(p) = \sum_{i=1}^{m} \frac{A_i}{p-\alpha_i}$$

where

$$A_i = [W(p)(p-\alpha_i)]_{p=\alpha_i},$$

and hence

$$w(t) = u(t) \sum_{i=1}^{m} A_i \, e^{\alpha_i t}.$$

On the other hand, if α_1 is repeated r_1 times; α_2, r_2 times; …; α_q, r_q times (where $r_1 + r_2 + \ldots + r_q = m$); the transfer function has the partial fraction expansion

$$W(p) = \sum_{i=1}^{q} \sum_{j=1}^{r_i} \frac{A_{ij}}{(p-\alpha_i)^j}$$

where

$$A_{ij} = \frac{1}{(r_i - j)!} \left[\left(\frac{d}{dp}\right)^{r_i - j} \left\{ W(p)(p-\alpha_i)^{r_i} \right\} \right]_{p=\alpha_i}$$

and hence

$$w(t) = u(t) \sum_{i=1}^{q} \sum_{j=1}^{r_i} \frac{A_{ij}}{(j-1)!} t^{j-1} e^{\alpha_i t}.$$

Relabelling the β_j's if necessary, we may suppose that they satisfy

$$\operatorname{Re}\beta_1 \leqslant \operatorname{Re}\beta_2 \leqslant \ldots \leqslant \operatorname{Re}\beta_h < 0 < \operatorname{Re}\beta_{h+1} \leqslant \operatorname{Re}\beta_{h+2} \leqslant \ldots \leqslant \operatorname{Re}\beta_n.$$

The inverse system has transfer function

$$W^{-1}(p) = \frac{D(p)}{N(p)} = c_{m-n}p^{m-n} + \ldots + c_2 p^2 + c_1 p + c_0 + \frac{C(p)}{N(p)}$$

where the $c_0, c_1, \ldots, c_{m-n}$ are constants and $C(p)$ is a polynomial in p of most degree $n-1$. Now $C(p)/N(p)$ is the transfer function of a dissipative system if and only if we choose the strip of convergence of $W^{-1}(p)$ to include the $i\omega$-axis; if and only if we choose the strip of convergence to be $\operatorname{Re}\beta_h < \operatorname{Re}p < \operatorname{Re}\beta_{h+1}$. We shall always make this choice. If $\beta_1, \beta_2, \ldots, \beta_n$ are distinct, then we have the partial fraction expansion

$$\frac{C(p)}{N(p)} = \sum_{j=1}^{n} \frac{C_j}{p-\beta_j}, \quad C_j = \left[\frac{C(p)}{N(p)}(p-\beta_j)\right]_{p=\beta_j}.$$

Hence the *inverse wavelet* is

$$w^{-1}(t) = c_{m-n}\,\delta^{(m-n)}(t) + \ldots + c_2\,\delta''(t) + c_1\,\delta'(t) + c_0\,\delta(t)$$
$$+ u(t)\sum_{j=1}^{h} C_j\, e^{\beta_j t} - u(-t)\sum_{j=h+1}^{n} C_j\, e^{\beta_j t}.$$

If some or all of the β_j's are repeated, we can find the expression for $w^{-1}(t)$ by the same method. Clearly, we see that $w^{-1}(t)$ is *realizable* if and only if no β_j has positive real part; that is, if and only if $w(t)$ is minimum-delay.

7.3. Inverse of a minimum-delay wavelet

Theorem 7.30. Let $w(t)$ be a non-null (i.e. invertible) wavelet and let $v(t)$ be either a null or non-null wavelet. Let their cross-product $w(t) \otimes v(t)$ be a one-sided time-function, i.e.

$$\int_0^\infty w(s+t)\,\overline{v(s)}\,ds = 0 \quad \text{for} \quad t<0.$$

Then $v(t) \equiv 0$ if and only if $w^{-1}(t)$ is a one-sided time-function.

Proof. Let $f(t) = w(t) \otimes v(t)$ and let $t \leqslant 0$. We have

$$\begin{aligned}
\overline{v(-t)} &= \int_0^\infty ds\,\delta(s+t)\,\overline{v(s)} \\
&= \int_0^\infty ds \int_{-\infty}^\infty dr\, w^{-1}(r)\, w(s+t-r)\,\overline{v(s)} \\
&= \int_{-\infty}^\infty dr\, w^{-1}(r)\left[\int_0^\infty ds\, w(s+t-r)\,\overline{v(s)}\right] \\
&= \int_{-\infty}^\infty dr\, w^{-1}(r) f(t-r) \\
&= \int_{-\infty}^t dr\, w^{-1}(r) f(t-r) \quad \text{(since } f(t) \text{ is one-sided).}
\end{aligned}$$

Hence $\overline{v(-t)} \equiv 0$ if and only if this last integral identically vanishes. Since $t < 0$, this last integral only involves $w^{-1}(r)$ for $r \leqslant t < 0$, and so this last integral identically vanishes if and only if $w^{-1}(r)$ is one-sided. Now let $t > 0$. Then $\overline{v(-t)} \equiv 0$ since $v(t)$ is a wavelet. Thus (for all t), $v(t) \equiv 0$ if and only if $w^{-1}(t)$ is one-sided. Q.E.D.

Theorems 6.30 and 7.30 immediately give the basic

Theorem 7.31. A wavelet $w(t)$ has a realizable inverse $w^{-1}(t)$ if and only if $w(t)$ is minimum-delay.

7.4. The first minimum-delay wavelet theorem

We now have our first main theorem on minimum-delay.

Theorem 7.40 (*the first minimum-delay wavelet theorem*). Let $w(t)$ be an invertible wavelet. Then each of the following conditions is necessary and sufficient that $w(t)$ is minimum-delay.

(a) The inverse wavelet $w^{-1}(t)$ is *realizable* (i.e. one-sided).

(b) The *group-delay* of the wavelet $w(t)$ is a *minimum* over the entire frequency range $(-\infty < \omega < \infty)$.

(c) The group-delay of the inverse wavelet $w^{-1}(t)$ is a *maximum* over the entire frequency range $(-\infty < \omega < \infty)$.

(d) The modulus

$$\left| \int_0^\infty w(t)\, \mathrm{e}^{-pt}\, dt \right|$$

of the transfer function of the wavelet $w(t)$ is a *maximum* in the right half-plane $(\mathrm{Re}\, p > 0)$.

(e) The modulus

$$\left| \int_{-\infty}^\infty w^{-1}(t)\, \mathrm{e}^{-pt}\, dt \right|$$

of the transfer function of the inverse wavelet is a *minimum* in the right half-plane $(\mathrm{Re}\, p > 0)$.

(f) The partial energy

$$\int_{-\infty}^{\tau} |w(t)|^2\, dt = \int_0^{\tau} |w(t)|^2\, dt$$

of the wavelet $w(t)$ is a *maximum* for all positive time $(\tau > 0)$.

(g) The partial energy

$$\int_{-\infty}^{\tau} |w^{-1}(t)|^2\, dt$$

of the inverse wavelet $w^{-1}(t)$ is a *minimum* for all negative time $(\tau < 0)$, and this minimum is zero.

(h) Any wavelet $v(t)$ such that

$$\int_0^\infty w(s+t)\,\overline{v(s)}\,ds = 0 \quad \text{for} \quad t<0$$

satisfies $v(t) \equiv 0$.

Note: *Maxima* and *minima* are with respect to wavelets in the class of all wavelets with the same gain as $w(t)$.

Proof. (a) Condition (a) follows from Theorem 7.31. Q.E.D. (a)

(b) By the canonical representation, every invertible wavelet $w(t)$ in the class of all wavelets with the same gain may be written

$$w(t) = a(t) * w_0(t)$$

where $a(t)$ is all-pass and $w_0(t)$ is the central minimum-delay wavelet for this gain. In terms of transfer functions, the canonical representation is

$$W(p) = A(p)\,W_0(p).$$

Let the notation $\tau_g(W)$ denote the group-delay of the transfer function $W(p)$. Then

$$\tau_g(W) = \tau_g(A) + \tau_g(W_0).$$

By Theorems 5.50 and 5.51, we have

$$\tau_g(A) \geqslant 0, \text{ and } = 0 \text{ if and only if } A \text{ is trivial.}$$

Hence $\tau_g(W)$ is a minimum if and only if $\tau_g(A) = 0$; i.e. if and only if $W(p)$ is minimum-delay. Q.E.D. (b).

(c) Condition (c) follows from condition (b) by noting that

$$\tau_g(W) = -\tau_g(W^{-1}). \qquad \text{Q.E.D. (c)}$$

(d) We have

$$|W(p)| = |A(p)|\,|W_0(p)|.$$

By Theorem 5.22, we have (in the right half-plane)

$$|A(p)| \leqslant 1, \text{ and } = 1 \text{ if and only if } A(p) \text{ is trivial.}$$

Hence $|W(p)|$ is a maximum if and only if $A(p)$ is trivial; i.e. if and only if $W(p)$ is minimum-delay. Q.E.D. (d).

(e) Condition (e) follows from condition (d) by noting that

$$w^{-1}(t) \leftrightarrow \int_{-\infty}^{\infty} w^{-1}(t)\,e^{-pt}\,dt = W^{-1}(p) = \frac{1}{W(p)}$$

so

$$|W^{-1}(p)| = |W(p)|^{-1}. \qquad \text{Q.E.D. (e)}$$

(f) Because $w(t) = a(t) * w_0(t)$, it follows from Theorem 5.60 that

$$\int_0^\tau |w(t)|^2\,dt \leqslant \int_0^\tau |w_0(t)|^2\,dt$$

with the equality sign holding for all τ if and only if $a(t)$ is trivial. That is, the partial energy of $w(t)$ is a maximum if and only if $a(t)$ is trivial; i.e. if and only if $w(t)$ is minimum-delay. Q.E.D. (f).

(g) The partial energy (for $\tau < 0$) of the inverse wavelet $w^{-1}(t)$ is zero if and only if $w^{-1}(t)$ is one-sided; i.e. if and only if $w(t)$ is minimum-delay. Q.E.D. (g).

(h) Condition (h) follows from the Karhunen Theorem 6.30. Q.E.D. (h).

CHAPTER VIII

LEAST-SQUARES AND TWO-STAGE LEAST-SQUARES

8.1. Matrices

Boldface letters, e.g. $\boldsymbol{\alpha}$, $\mathbf{x}$, will denote *matrices* with complex entries, e.g. α_{ij}, x_i. Matrices are rectangular (i.e. n by m) arrays, and, in particular, they may be *square* (i.e. n by n), or *column vectors* (i.e. n by 1), for example:

$$\boldsymbol{\alpha} = [\alpha_{ij}], \quad i,j = 1,2,\ldots,n;$$

$$\mathbf{x} = [x_i], \quad i = 1,2,\ldots,n \quad \text{(column vector).}$$

A bar over a matrix will indicate its *adjoint* (i.e. its complex-conjugate transpose); i.e.

$$\bar{\boldsymbol{\alpha}} = [\bar{\alpha}_{ji}], \quad i,j = 1,2,\ldots,n; \qquad \bar{\mathbf{x}} = (\bar{x}_1,\ldots,\bar{x}_n) \qquad \text{(row vector).}$$

(Note: As usual, a bar over a complex number indicates its complex-conjugate; i.e. $\bar{\alpha}_{ji}$ is the complex-conjugate of α_{ji}.)

The *null matrix* $\mathbf{0}$ is a matrix whose entries are zeros. The *identity matrix* $\mathbf{I}$ is the square matrix with ones down the main diagonal and zeros off the diagonal. A *hermitian matrix* $\boldsymbol{\Phi}$ is a square matrix that is equal to its adjoint; i.e. $\boldsymbol{\Phi} = \bar{\boldsymbol{\Phi}}$. A *unitary matrix* $\mathbf{A}_0$ is a square matrix whose inverse and adjoint are equal, i.e. $\mathbf{A}_0^{-1} = \bar{\mathbf{A}}_0$ or $\mathbf{A}_0 \bar{\mathbf{A}}_0 = \mathbf{I}$.

Matrix addition and multiplication are defined as usual; i.e.

$$\boldsymbol{\alpha} + \boldsymbol{\beta} = [\alpha_{ij} + \beta_{ij}], \quad \boldsymbol{\alpha\beta} = \left[\sum_{k=1}^{n} \alpha_{ik}\beta_{kj}\right].$$

Matrix multiplication is not necessarily commutative.

Vertical bars enclosing a matrix are defined by

$$|\boldsymbol{\alpha}|^2 = \boldsymbol{\alpha}\bar{\boldsymbol{\alpha}}, \quad |\mathbf{x}|^2 = \mathbf{x}\bar{\mathbf{x}}.$$

We see that $|\boldsymbol{\alpha}|^2$ and $|\mathbf{x}|^2$ are hermitian matrices. (Note: As usual, $|\alpha_{ij}|$ denotes the absolute value of the complex number α_{ij}.)

The *determinant* of a square matrix $\boldsymbol{\alpha}$ is denoted by $\det \boldsymbol{\alpha}$. The *inverse* of a square matrix $\boldsymbol{\alpha}$ is denoted by $\boldsymbol{\alpha}^{-1}$; i.e. $\boldsymbol{\alpha\alpha}^{-1} = \boldsymbol{\alpha}^{-1}\boldsymbol{\alpha} = \mathbf{I}$.

The *trace* of a square matrix $\boldsymbol{\alpha}$ is defined to be the sum of its entries down the main diagonal, and is denoted by $\operatorname{tr}\boldsymbol{\alpha}$; i.e.

$$\operatorname{tr}\boldsymbol{\alpha} = \sum_{i=1}^{n} \alpha_{ii} = \alpha_{11} + \ldots + \alpha_{nn}.$$

8.2. Linear matrix space

Let $\mathbf{L}^2$ denote an *abstract space* with *elements* $\mathbf{f}, \mathbf{g}, \mathbf{h}$. These elements may be n by m rectangular matrices. Let the *constants* $\boldsymbol{\alpha}, \boldsymbol{\beta}$ be n by n square matrices with complex entries. Then we see that addition $\mathbf{f}+\mathbf{g}$, and multiplication $\boldsymbol{\alpha}\mathbf{f}$ by a constant, are defined.

Postulate P_1. The space $\mathbf{L}^2$ is linear.

Note: For the definition of *linear space*, see Robinson (1959, p. 44), except that constants are now square matrices instead of scalars, as given there.

Definition 8.20. If a constant $(\mathbf{f}, \mathbf{g})$ exists for each pair of elements $\mathbf{f}, \mathbf{g}$ of linear space such that

(1) $(\boldsymbol{\alpha}\mathbf{f}, \mathbf{g}) = \boldsymbol{\alpha}(\mathbf{f}, \mathbf{g})$ where $\boldsymbol{\alpha}$ is a constant,

(2) $(\mathbf{f}+\mathbf{g}, \mathbf{h}) = (\mathbf{f}, \mathbf{h}) + (\mathbf{g}, \mathbf{h})$,

(3) $(\mathbf{f}, \mathbf{g}) = \overline{(\mathbf{g}, \mathbf{f})}$,

(4) $\operatorname{tr}(\mathbf{f}, \mathbf{f}) \geqslant 0$, and $= 0$ if and only if $\mathbf{f} = 0$,

then $(\mathbf{f}, \mathbf{g})$ is called the *Gramian* of $\mathbf{f}$ and $\mathbf{g}$.

Postulate P_2. A Gramian is defined for the linear space $\mathbf{L}^2$.

Condition (3) states that the Gramian is a hermitian matrix. Moreover, it follows that

(5) $(\mathbf{f}, \boldsymbol{\alpha}\mathbf{g}) = (\mathbf{f}, \mathbf{g})\bar{\boldsymbol{\alpha}}$,

(6) $(\mathbf{f}, \mathbf{g}+\mathbf{h}) = (\mathbf{f}, \mathbf{g}) + (\mathbf{f}, \mathbf{h})$.

The positive square root of $\operatorname{tr}(\mathbf{f}, \mathbf{f})$ is called the *norm* of $\mathbf{f}$ and is denoted by $\|\mathbf{f}\|$.

Postulate P_3. The linear space $\mathbf{L}^2$, with norm $\|\mathbf{f}\|$, is complete. We shall assume that space $\mathbf{L}^2$ satisfies Postulates P_1, P_2, and P_3.

Two elements $\mathbf{f}$ and $\mathbf{g}$ are called *normal*, or *perpendicular*, or *orthogonal*, when $(\mathbf{f}, \mathbf{g}) = \mathbf{0}$, in which case we write $\mathbf{f} \perp \mathbf{g}$. If $\mathbf{f} \perp \mathbf{g}$, then the Pythagorean theorem states that

$$\|\mathbf{f}\|^2 + \|\mathbf{g}\|^2 = \|\mathbf{f}+\mathbf{g}\|^2.$$

A sequence $\mathbf{f}_N$ $(N = 0, 1, 2, \ldots)$ of elements is said to *converge in the mean* to an element $\mathbf{f}$ if the numbers $\|\mathbf{f} - \mathbf{f}_N\|$ converge to zero as $N \to \infty$. We then write

$$\mathbf{f}_N \Rightarrow \mathbf{f} \quad \text{or} \quad \underset{N\to\infty}{\text{l.i.m.}}\, \mathbf{f}_N = \mathbf{f}.$$

where l.i.m. stands for *limit in the mean*.

Definition 8.21. A set $\{\mathbf{g}(r), r \in R\}$ of elements in $\mathbf{L}^2$ is said to be *maximal* in a subspace of $\mathbf{L}^2$, or to *span* a subspace of $\mathbf{L}^2$, if for every element $\mathbf{f}$ of the subspace there exist constants $\boldsymbol{\lambda}_{Nj}$ and elements $\mathbf{g}(r_{Nj})$ of the set such that

$$\sum_{j=1}^{N} \boldsymbol{\lambda}_{Nj}\, \mathbf{g}(r_{Nj}) \Rightarrow \mathbf{f} \quad \text{as} \quad N \to \infty.$$

Instead of the above analytic definition, we may use the operational definition:

Definition 8.22. A set $\{\mathbf{g}(r), r \in R\}$ of elements in $\mathbf{L}^2$ is said to be *maximal* in a subspace of $\mathbf{L}^2$, or to *span* a subspace of $\mathbf{L}^2$, if for every element $\mathbf{f}$ of the subspace there exists a *generalized function* $\boldsymbol{\lambda}(r)$ such that

$$\int_{r \in R} \boldsymbol{\lambda}(r)\, \mathbf{g}(r)\, dr = \mathbf{f}.$$

The *subspace spanned by the set* $\{\mathbf{g}(r), r \in R\}$ is denoted by:

Subspace spanned by $\{\mathbf{g}(r), r \in R\}$.

The notation "Subspace spanned by" may be abbreviated to "S.s.b.".

The following theorem is the $\mathbf{L}^2$-space counterpart of a well-known theorem for Hilbert space.

Theorem 8.20. A set $\{\mathbf{g}(r), r \in R\}$ is maximal in a subspace of $\mathbf{L}^2$ if and only if every element $\mathbf{f}$ of the subspace such that

$$\mathbf{f} \perp \mathbf{g}(r) \quad \text{for all} \quad r \in R$$

identically vanishes (i.e. $\mathbf{f} \equiv 0$).

8.3. The method of least-squares

In this section, we let:

$\mathbf{f}$ be an element of $\mathbf{L}^2$: $\mathbf{f} \in \mathbf{L}^2$,
$\mathbf{S}_1$ be a subspace of $\mathbf{L}^2$: $\mathbf{S}_1 \subset \mathbf{L}^2$,
$\mathbf{f}_1$ be an element of $\mathbf{S}_1$: $\mathbf{f}_1 \in \mathbf{S}_1$.

Definition 8.30. $\mathbf{f}_1$ is called the *least-squares approximation* to $\mathbf{f}$ if the *mean-square error*

$$I = \|\mathbf{f} - \mathbf{f}_1\|^2$$

is a minimum. The element $\mathbf{f}_1$ is then said to be determined by the *method of least-squares*, and $\mathbf{f} - \mathbf{f}_1$ is called the *least-squares error*.

The following theorem is the $\mathbf{L}^2$-space counterpart of the projection theorem for Hilbert space (see Robinson, 1959, p. 55).

Theorem 8.30. $\mathbf{f}_1$ is the least-squares approximation to $\mathbf{f}$ if and only if the error $\mathbf{f}-\mathbf{f}_1$ is normal to $\mathbf{S}_1$.

Theorem 8.31. If $\mathbf{f}_1$ is the least-squares approximation to $\mathbf{f}$, then:

(a) $\mathbf{f}-\mathbf{f}_1 \perp \mathbf{f}_1$.

(b) $\|\mathbf{f}\|^2 = \|\mathbf{f}_1\|^2 + \|\mathbf{f}-\mathbf{f}_1\|^2$.

(c) $(\mathbf{f}, \mathbf{f}_1) = (\mathbf{f}_1, \mathbf{f}) = (\mathbf{f}_1, \mathbf{f}_1)$.

(d) The minimum mean-square error is

$$I_{\min} = \|\mathbf{f}-\mathbf{f}_1\|^2 = \|\mathbf{f}\|^2 - \|\mathbf{f}_1\|^2 = \|\mathbf{f}\|^2 - \operatorname{tr}(\mathbf{f}, \mathbf{f}_1).$$

Proof. Because $\mathbf{f}_1$ is an element of $\mathbf{S}_1$, condition (a) follows from Theorem 8.30. Condition (b) follows from condition (a) by the Pythagorean theorem. Condition (c) follows from

$$(\mathbf{f}, \mathbf{f}_1) = (\mathbf{f}-\mathbf{f}_1+\mathbf{f}_1, \mathbf{f}_1) = (\mathbf{f}_1, \mathbf{f}_1) = \overline{(\mathbf{f}_1, \mathbf{f}_1)} = (\mathbf{f}_1, \mathbf{f}).$$

Condition (d) follows from conditions (b) and (c). Q.E.D.

From Theorem 8.30, we obtain

Theorem 8.32. Let the set $\{\mathbf{g}(r), r \in R\}$ span $\mathbf{S}_1$. Then $\mathbf{f}_1$ is the least-squares approximation to $\mathbf{f}$ if and only if $\mathbf{f}-\mathbf{f}_1 \perp \mathbf{g}(r)$ for $r \in R$.

That is, $\mathbf{f}_1$ is determined by the method of least-squares if and only if the equations

$$(\mathbf{f}-\mathbf{f}_1, \mathbf{g}(r)) = 0$$

hold for $r \in R$. Since these equations express that the error $\mathbf{f}-\mathbf{f}_1$ be normal, or orthogonal, to each $\mathbf{g}(r)$ (where $r \in R$), they are called the *normal equations.*

8.4. Two-stage least squares

In this section, we let:

$\mathbf{f}$ be an element of $\mathbf{L}^2$: $\mathbf{f} \in \mathbf{L}^2$,

$\mathbf{S}_1$ be a subspace of $\mathbf{L}^2$: $\mathbf{S}_1 \subset \mathbf{L}^2$,

$\mathbf{f}_1$ be an element of $\mathbf{S}_1$: $\mathbf{f}_1 \in \mathbf{S}_1$,

$\mathbf{S}_2$ be a subspace of $\mathbf{S}_1$: $\mathbf{S}_2 \subset \mathbf{S}_1$,

$\mathbf{f}_2$ be an element of $\mathbf{S}_2$: $\mathbf{f}_2 \in \mathbf{S}_2$.

Theorem 8.40 (the two-stage least-squares theorem). If $\mathbf{f}_1$ is the least-squares approximation to $\mathbf{f}$, and if $\mathbf{f}_2$ is the least-squares approximation to $\mathbf{f}_1$, then $\mathbf{f}_2$ is the least-squares approximation to $\mathbf{f}$, with (minimum) mean-square error

$$I_{\min} = \|\mathbf{f}-\mathbf{f}_2\|^2 = \|\mathbf{f}-\mathbf{f}_1\|^2 + \|\mathbf{f}_1-\mathbf{f}_2\|^2.$$

Proof. By hypothesis, $\mathbf{f}-\mathbf{f}_1 \perp \mathbf{S}_1$. But $\mathbf{S}_1 \supset \mathbf{S}_2$. Hence $\mathbf{f}-\mathbf{f}_1 \perp \mathbf{S}_2$. Also by hypothesis, $\mathbf{f}_1-\mathbf{f}_2 \perp \mathbf{S}_2$. Therefore

$$(\mathbf{f}-\mathbf{f}_1)+(\mathbf{f}_1-\mathbf{f}_2) = \mathbf{f}-\mathbf{f}_2 \perp \mathbf{S}_2,$$

so $\mathbf{f}_2$ is the least-squares approximation to $\mathbf{f}$. Since $\mathbf{f}_1 \in \mathbf{S}_1$ and $\mathbf{f}_2 \in \mathbf{S}_2 \subset \mathbf{S}_1$, it follows that $\mathbf{f}_1-\mathbf{f}_2 \in \mathbf{S}_1$. But $\mathbf{f}-\mathbf{f}_1 \perp \mathbf{S}_1$. Therefore

$$\mathbf{f}-\mathbf{f}_1 \perp \mathbf{f}_1-\mathbf{f}_2. \qquad \text{Q.E.D.}$$

8.5. Matrix-valued functions

The functions appearing in the first seven chapters are scalar-valued. For example, the time-function $f(t)$ assumes values that are complex numbers (i.e. scalars). We now wish to introduce ***matrix-valued functions.*** For example, the (matrix-valued) time-function $\mathbf{f}(t)$ assumes values that are matrices whose entries are complex numbers. Alternatively, we may say that $\mathbf{f}(t)$ is a matrix whose entries are scalar-valued time-functions $f_{ij}(t)$; i.e.

$$\mathbf{f}(t) = [f_{ij}(t)], \quad i,j = 1,2,\ldots,n.$$

The integral of a matrix-valued function $\mathbf{f}(t)$ is defined to be the matrix obtained by integrating each entry $f_{ij}(t)$; i.e.

$$\int_{-\infty}^{\infty} \mathbf{f}(t)\,dt = \left[\int_{-\infty}^{\infty} f_{ij}(t)\,dt\right], \quad i,j = 1,2,\ldots,n.$$

The following spaces are specific realizations of abstract space $\mathbf{L}^2$.

(a) Space $\mathbf{L}_t^2$: the space of all *L^2-stable* (*square-matrix-valued*) *time-functions.*

Elements: all $\mathbf{f}(t)$ such that $\operatorname{tr}\int_{-\infty}^{\infty} |\mathbf{f}(t)|^2\,dt < \infty$.

Gramian: $(\mathbf{f},\mathbf{g})_t = \int_{-\infty}^{\infty} \mathbf{f}(t)\,\overline{\mathbf{g}(t)}\,dt.$

Norm: $\|\mathbf{f}\|_t = \sqrt{[\operatorname{tr}(\mathbf{f},\mathbf{f})_t]}$.

(b) Space $\mathbf{L}_t^2(0,\infty)$: (*square-matrix-valued*) *wavelet space.* This space is a proper subspace of $\mathbf{L}_t^2$.

Elements: all $\mathbf{w}(t)$ such that $\mathbf{w}(t) = 0$ for $t<0$ and

$\operatorname{tr}\int_0^{\infty} |\mathbf{w}(t)|^2\,dt < \infty$. (Such $\mathbf{w}(t)$ are called *wavelets.*)

Gramian: same as that for $\mathbf{L}_t^2$.

Norm: same as that for $\mathbf{L}_t^2$.

(c) Space $\mathbf{L}^2_\omega$: *the space of all L^2-stable (square-matrix-valued) frequency-functions.*

Elements: all $\mathbf{F}(i\omega)$ such that $\operatorname{tr}\frac{1}{2\pi}\int_{-\infty}^{\infty}|\mathbf{F}(i\omega)|^2\,d\omega<\infty$.

Gramian: $(\mathbf{F},\mathbf{G})_\omega=\frac{1}{2\pi}\int_{-\infty}^{\infty}\mathbf{F}(i\omega)\,\overline{\mathbf{G}(i\omega)}\,d\omega$.

Norm: $\|\mathbf{F}\|_\omega=\sqrt{[\operatorname{tr}(\mathbf{F},\mathbf{F})_\omega]}$.

(d) Space $\mathbf{L}^2_P$: the space of all (matrix-valued) variates with finite second absolute moments (see Section **10.1**).

Elements: all variates $\mathbf{x}(\xi)$ such that

$$\operatorname{tr}\int_{\text{all }\xi}|\mathbf{x}(\xi)|^2\,P(d\xi)<\infty.$$

Gramian: $(\mathbf{x},\mathbf{y})_P=((\mathbf{x},\mathbf{y}))=\int_{\text{all }\xi}\mathbf{x}(\xi)\,\overline{\mathbf{y}(\xi)}\,P(d\xi)$.

Norm: $\|\mathbf{x}\|_P=\|\mathbf{x}\|=\sqrt{[\operatorname{tr}((\mathbf{x},\mathbf{x}))]}$.

(*Remarks on notation*: The time-variable t appearing in the Gramian $(\mathbf{f},\mathbf{g})_t$ is a dummy variable of integration, and hence any other time-variable, e.g. r, s, τ, may be used equally as well for the subscript of the Gramian; i.e.

$$(\mathbf{f}(t),\mathbf{g}(t))_t=(\mathbf{f}(s),\mathbf{g}(s))_s=\int_{-\infty}^{\infty}\mathbf{f}(s)\,\overline{\mathbf{g}(s)}\,ds.$$

This same remark applies to the norm $\|\mathbf{f}\|_t$, and the corresponding remark holds for the Gramian and norm of $\mathbf{L}^2_\omega$. Because we make extensive use of space $\mathbf{L}^2_P$ in Chapters X and XI, we give its notation special treatment; namely, in those chapters, we let double parentheses with no subscript denote the Gramian, and double vertical bars with no subscript denote the norm, for space $\mathbf{L}^2_P$.)

We shall let $\mathbf{f}(t)\leftrightarrow\mathbf{F}(i\omega)$ denote:

$$\mathbf{f}(t)=\frac{1}{2\pi}\int_{-\infty}^{\infty}e^{i\omega t}\mathbf{F}(i\omega)\,d\omega\quad\text{and}\quad\mathbf{F}(i\omega)=\int_{-\infty}^{\infty}e^{-i\omega t}\mathbf{f}(t)\,dt.$$

If $\mathbf{f}(t)\leftrightarrow\mathbf{F}(i\omega)$ and $\mathbf{g}(t)\leftrightarrow\mathbf{G}(i\omega)$ are L^2-stable, then we have *Parseval's equality*

$$\int_{-\infty}^{\infty}\mathbf{f}(t)\,\overline{\mathbf{g}(t)}\,dt=\frac{1}{2\pi}\int_{-\infty}^{\infty}\mathbf{F}(i\omega)\,\overline{\mathbf{G}(i\omega)}\,d\omega$$

or, in terms of Gramians,

$$(\mathbf{f},\mathbf{g})_t=(\mathbf{F},\mathbf{G})_\omega.$$

CHAPTER IX

SYSTEMS WITH MULTIPLE INPUTS AND OUTPUTS

9.1. Multiple systems

We now consider *cause–effect time-invariant linear systems with multiple inputs and outputs.* We have

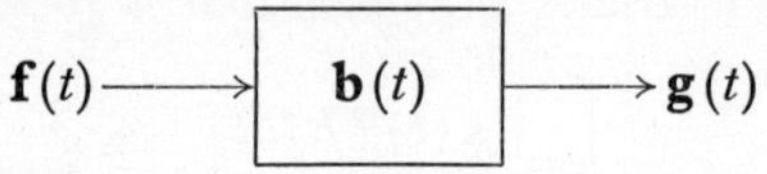

where the input $\mathbf{f}(t)$, impulse response $\mathbf{b}(t)$, and the output $\mathbf{g}(t)$ are matrix-valued time-functions. The impulse response $\mathbf{b}(t)$ is a square matrix, but both the input and output can be rectangular matrices. Input and output are related by

$$\int_{-\infty}^{\infty} \mathbf{b}(s)\,\mathbf{f}(t-s)\,ds = \mathbf{g}(t)$$

which we write operationally as

$$\mathbf{b} * \mathbf{f} = \mathbf{g}.$$

In terms of fourier transforms, we have

$$\mathbf{BF} = \mathbf{G}.$$

The *square-gain* of the system $\mathbf{b}(t) \leftrightarrow \mathbf{B}(p)$ is defined to be

$$\mathbf{\Phi}(i\omega) = \mathbf{M}^2(i\omega) = \mathbf{B}(i\omega)\,\overline{\mathbf{B}(i\omega)} = |\mathbf{B}(i\omega)|^2.$$

Definition 9.10. A system $\mathbf{b}(t) \leftrightarrow \mathbf{B}(i\omega)$ is said to be *invertible* if the inverse matrix $\mathbf{B}^{-1}(i\omega)$ exists for almost all ω. Then the impulse response $\mathbf{b}^{-1}(t)$ of the *inverse system* is defined by

$$\mathbf{b}^{-1}(t) \leftrightarrow \mathbf{B}^{-1}(i\omega).$$

Since

$$\mathbf{B}^{-1}(i\omega)\,\mathbf{B}(i\omega) = \mathbf{B}(i\omega)\,\mathbf{B}^{-1}(i\omega) = \mathbf{I},$$

we have

$$\mathbf{b}^{-1}(t) * \mathbf{b}(t) = \mathbf{b}(t) * \mathbf{b}^{-1}(t) = \boldsymbol{\delta}(t)$$

where $\boldsymbol{\delta}(t)$ is defined to be

$$\boldsymbol{\delta}(t) = \delta(t)\,\mathbf{I}.$$

9.2. Matrix-valued wavelets

A system is called *dissipative* if $\mathbf{b}(t) \leftrightarrow \mathbf{B}(i\omega)$ is L^2-stable. A system is called *realizable* if $\mathbf{b}(t) = 0$ for $t<0$ (i.e. if $\mathbf{b}(t)$ is *one-sided*). Thus, a system is realizable and dissipative if and only if its impulse response is a wavelet.

Definition 9.20. A wavelet (or realizable, dissipative system) $\mathbf{w}(t)$ is called *minimum-delay* if it has a realizable (i.e. one-sided) inverse $\mathbf{w}^{-1}(t)$.

9.3. Auto-products and cross-products of matrix-valued time-functions

Definition 9.30. The *cross-product* $\boldsymbol{\phi}_{\mathbf{fg}}(t) \equiv \mathbf{f}(t) \otimes \mathbf{g}(t)$ of the time-function $\mathbf{f}(t)$ with the time-function $\mathbf{g}(t)$ is defined to be

$$\boldsymbol{\phi}_{\mathbf{fg}}(t) = \mathbf{f}(t) * \overline{\mathbf{g}(-t)} = \int_{-\infty}^{\infty} \mathbf{f}(s+t)\,\overline{\mathbf{g}(s)}\,ds = (\mathbf{f}(s+t), \mathbf{g}(s))_s.$$

If $\mathbf{f}(t) = \mathbf{g}(t)$, then $\boldsymbol{\phi}_{\mathbf{ff}}(t)$ is called the *auto-product* of $\mathbf{f}(t)$.

If $\mathbf{f}(t) \leftrightarrow \mathbf{F}(i\omega)$, then $\overline{\mathbf{f}(-t)} \leftrightarrow \overline{\mathbf{F}(i\omega)}$. Hence

$$\boldsymbol{\phi}_{\mathbf{ff}}(t) = \mathbf{f}(t) * \overline{\mathbf{f}(-t)} \leftrightarrow \boldsymbol{\Phi}_{\mathbf{ff}}(i\omega) = \mathbf{F}(i\omega)\overline{\mathbf{F}(i\omega)}$$

and

$$\boldsymbol{\phi}_{\mathbf{fg}}(t) = \mathbf{f}(t) * \overline{\mathbf{g}(-t)} \leftrightarrow \boldsymbol{\Phi}_{\mathbf{fg}}(i\omega) = \mathbf{F}(i\omega)\overline{\mathbf{G}(i\omega)}.$$

We also see that

$$\overline{\boldsymbol{\phi}_{\mathbf{ff}}(-t)} = \boldsymbol{\phi}_{\mathbf{ff}}(t) \leftrightarrow \overline{\mathbf{F}(i\omega)\overline{\mathbf{F}(i\omega)}} = \mathbf{F}(i\omega)\overline{\mathbf{F}(i\omega)}$$

and

$$\overline{\boldsymbol{\phi}_{\mathbf{fg}}(-t)} = \boldsymbol{\phi}_{\mathbf{gf}}(t) \leftrightarrow \overline{\mathbf{F}(i\omega)\overline{\mathbf{G}(i\omega)}} = \mathbf{G}(i\omega)\overline{\mathbf{F}(i\omega)}.$$

Moreover, two L^2-stable time-functions $\mathbf{f}_1(t)$ and $\mathbf{f}_2(t)$ each have the same square-gain if and only if they each have the same autoproduct; i.e.

$$\mathbf{F}_1\overline{\mathbf{F}_1} = \mathbf{F}_2\overline{\mathbf{F}_2} \quad \text{if and only if} \quad \boldsymbol{\phi}_{11}(t) = \boldsymbol{\phi}_{22}(t).$$

9.4. The second minimum-delay wavelet theorem

We now have our second main theorem on minimum-delay.

Theorem 9.40 (*the second minimum-delay wavelet theorem*). Let $\mathbf{w}(t)$ be an invertible wavelet. Then each of the following conditions is necessary and sufficient that $\mathbf{w}(t)$ is minimum-delay.

(a) Any wavelet $\mathbf{v}(t)$ such that

$$\int_0^{\infty} \mathbf{w}(s+t)\,\overline{\mathbf{v}(s)}\,ds = 0 \quad \text{for} \quad t<0$$

must identically vanish (i.e. $\mathbf{v}(t) \equiv 0$).

(b) The set $\{\mathbf{w}(t-r), r \geqslant 0\}$ is maximal in wavelet space $\mathbf{L}_t^2(0, \infty)$.

(c) For every wavelet $\mathbf{v}(t)$ there is a realizable $\gamma(t)$ with input $\mathbf{w}(t)$ and with output

$$\mathbf{v}(t) = \int_0^\infty \gamma(r)\,\mathbf{w}(t-r)\,dr.$$

(d) For every wavelet $\mathbf{v}(t)$ the system

$$\gamma(t) = \mathbf{v}(t) * \mathbf{w}^{-1}(t)$$

is realizable.

Proof. By Definition 9.20, $\mathbf{w}(t)$ is minimum-delay if and only if $\mathbf{w}^{-1}(t)$ is realizable (i.e. one-sided). Condition (a) now follows by the same proof as Theorem 7.30, except that matrix-valued functions replace scalar-valued functions. To prove condition (b) we observe that

$$0 = \int_0^\infty \mathbf{w}(s+t)\,\overline{\mathbf{v}(s)}\,ds = \int_0^\infty \mathbf{w}(s-r)\,\overline{\mathbf{v}(s)}\,ds$$

$$= (\mathbf{w}(s-r), \mathbf{v}(s))_s \quad (\text{for all } r = -t > 0)$$

means that the wavelet $\mathbf{v}(s)$ is orthogonal to the set $\{\mathbf{w}(s-r), r \geqslant 0\}$. By Theorem 8.20, the set is maximal if and only if these orthogonality relations imply that $\mathbf{v}(t)$ identically vanishes. But, by condition (a), $\mathbf{v}(t) \equiv 0$ if and only if $\mathbf{w}(t)$ is minimum-delay. Condition (b) is thereby established. Condition (c) is the operational definition (Definition 8.22) of a maximal set, and so this condition follows from condition (b). Condition (d) is a restatement of condition (c). Q.E.D.

9.5. Matrix-valued all-pass systems

Definition 9.50. A system $\mathbf{a}(t) \leftrightarrow \mathbf{A}(i\omega)$ is called an *all-pass system* if it is realizable, i.e.

$$\mathbf{a}(t) = 0 \quad \text{for} \quad t < 0,$$

and has unit square-gain, i.e.

$$|\mathbf{A}(i\omega)|^2 = \mathbf{I} \quad (\text{for almost all } \omega).$$

Since

$$\mathbf{A}(i\omega)\,\overline{\mathbf{A}(i\omega)} = \mathbf{I} \quad (\text{for almost all } \omega),$$

an all-pass system $\mathbf{A}(i\omega)$ is always invertible, and has the inverse

$$\mathbf{A}^{-1}(i\omega) = \overline{\mathbf{A}(i\omega)}.$$

Definition 9.51. An all-pass system $\mathbf{a}(t)$ is called *non-trivial* if its inverse $\mathbf{a}^{-1}(t)$ is non-realizable, and is called *trivial* if its inverse is realizable.

If $\mathbf{w}(t) \leftrightarrow \mathbf{W}(p)$ is a wavelet, and if $\mathbf{a}(t) \leftrightarrow \mathbf{A}(p)$ is all-pass, then

$$\mathbf{v}(t) = \mathbf{w}(t) * \mathbf{a}(t) \leftrightarrow \mathbf{V}(p) = \mathbf{W}(p)\,\mathbf{A}(p)$$

is a wavelet with the same square-gain

$$\mathbf{V}(i\omega)\,\overline{\mathbf{V}(i\omega)} = \mathbf{W}(i\omega)\,\mathbf{A}(i\omega)\,\overline{\mathbf{A}(i\omega)}\;\overline{\mathbf{W}(i\omega)} = \mathbf{W}(i\omega)\,\overline{\mathbf{W}(i\omega)}.$$

The matrix-valued counterpart of Theorem 5.60 is

Theorem 9.50 (*the partial energy theorem*). Let a wavelet $\mathbf{w}(t)$ be the input to an all-pass system $\mathbf{A}(p)$ and let the wavelet $\mathbf{v}(t)$ be the resulting output. Then the partial energy of input is greater than or equal to the partial energy of output, i.e.

$$\operatorname{tr}\int_0^\tau |\mathbf{w}(t)|^2\,dt \geqslant \operatorname{tr}\int_0^\tau |\mathbf{v}(t)|^2\,dt.$$

The all-pass system is trivial if and only if the equality holds for all τ. The all-pass system is non-trivial if and only if the strict inequality (i.e. *greater than*) holds for some $\tau > 0$. For $\tau = \infty$, the equality holds (i.e. the total energies are equal).

An all-pass system is trivial if and only if its transfer function is a constant, unitary matrix $\mathbf{A}_0$.

9.6. The canonical representation of matrix-valued wavelets

The matrix-valued counterparts of Theorems 6.10 and 6.11 are:

Theorem 9.60 (*the canonical representation*). $\mathbf{W}(p)$ is the transfer function of an invertible, realizable, dissipative system if and only if

$$\mathbf{W}(p) = \mathbf{W}_0(p)\,\mathbf{A}(p)$$

where $\mathbf{A}(p)$ is an all-pass transfer function and $\mathbf{W}_0(p)$ is a minimum-delay transfer function with the square-gain $|\mathbf{W}(i\omega)|^2$.

Theorem 9.61 (*the canonical representation*). $\mathbf{w}(t)$ is an invertible wavelet if and only if

$$\mathbf{w}(t) = \mathbf{w}_0(t) * \mathbf{a}(t)$$

where $\mathbf{a}(t)$ is an all-pass impulse response function and $\mathbf{w}_0(t)$ is a minimum-delay wavelet with the same square-gain as $\mathbf{w}(t)$.

The inverse of the invertible wavelet with canonical representation $\mathbf{W}(p) = \mathbf{W}_0(p)\,\mathbf{A}(p)$ is

$$\mathbf{W}^{-1}(p) = [\mathbf{W}_0(p)\,\mathbf{A}(p)]^{-1} = \mathbf{A}^{-1}(p)\,\mathbf{W}_0^{-1}(p).$$

This inverse is realizable (i.e. $\mathbf{W}(p)$ is minimum-delay) if and only if $\mathbf{A}^{-1}(p)$ is realizable (i.e. $\mathbf{A}(p) = \mathbf{A}_0$ is trivial). That is, two minimum-delay wavelets with the same square-gain are equal except for a trivial all-pass system $\mathbf{A}_0$. Any minimum-delay wavelet $\mathbf{W}_0(p)\,\mathbf{A}_0$ is determined (except for the trivial all-pass system $\mathbf{A}_0$) by its square-gain

$$\mathbf{\Phi}(i\omega) = \mathbf{W}_0(i\omega)\,\mathbf{A}_0\,\overline{[\mathbf{W}_0(i\omega)\,\mathbf{A}_0]} = \mathbf{W}_0(i\omega)\,\overline{\mathbf{W}_0(i\omega)}.$$

9.7. Approximation of an L^2-stable time-function by a wavelet

In this section we let:

$\mathbf{w}(t)$ be an invertible wavelet,
$\boldsymbol{\varkappa}(t)$ be a non-realizable system,
$\mathbf{h}(t) = \boldsymbol{\varkappa}(t) * \mathbf{w}(t)$ be an L^2-stable time-function,
$\boldsymbol{\gamma}(t)$ be a realizable system,
$\mathbf{v}(t) = \boldsymbol{\gamma}(t) * \mathbf{w}(t)$ be a wavelet.

That is, we have:

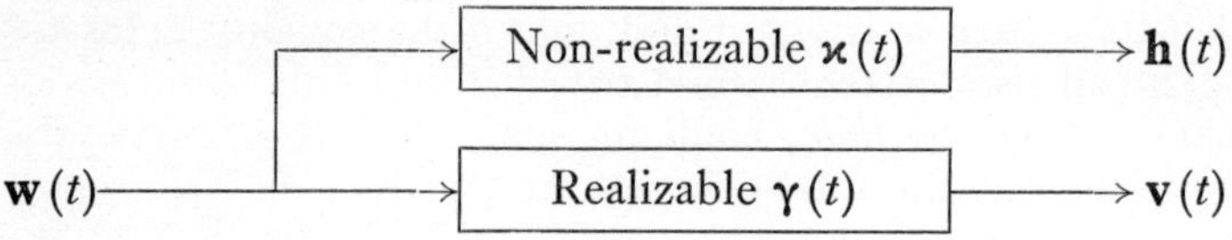

The wavelet $\mathbf{v}(t) = \boldsymbol{\gamma}(t) * \mathbf{w}(t)$ is in the subspace:

$$\mathbf{S} = \text{Subspace spanned by } \{\mathbf{w}(t-r), r \geqslant 0\}.$$

Clearly $\mathbf{S}$ is a subspace of wavelet space $\mathbf{L}_t^2(0,\infty)$; and by Theorem 9.40(b) $\mathbf{S}$ coincides with $\mathbf{L}_t^2(0,\infty)$ if and only if $\mathbf{w}(t)$ is minimum-delay.

Theorem 9.70. The wavelet $\boldsymbol{\gamma}(t) * \mathbf{w}(t)$ is the least-squares approximation to the L^2-stable time-function $\boldsymbol{\varkappa}(t) * \mathbf{w}(t)$ if and only if

$$\boldsymbol{\gamma}(t) = \{[\boldsymbol{\varkappa}(t) * \mathbf{w}_0(t)]\,u(t)\} * \mathbf{w}_0^{-1}(t),$$

$$\mathbf{\Gamma}(i\omega) = \left\{\int_0^\infty dt\,\mathrm{e}^{-i\omega t}\frac{1}{2\pi}\int_{-\infty}^\infty d\mu\,\mathrm{e}^{i\mu t}\,\mathbf{K}(i\mu)\,\mathbf{W}_0(i\mu)\right\}\mathbf{W}_0^{-1}(i\omega)$$

where $\mathbf{w}_0(t) \leftrightarrow \mathbf{W}_0(i\omega)$ is minimum-delay with the same square-gain as $\mathbf{w}(t)$.

Preliminary remarks. We wish to find the system $\boldsymbol{\gamma}(t)$ that minimizes

$$\|\mathbf{h}(t) - \mathbf{v}(t)\|_t^2 = \|\boldsymbol{\varkappa}(t) * \mathbf{w}(t) - \boldsymbol{\gamma}(t) * \mathbf{w}(t)\|_t^2. \tag{9.70}$$

Since by definition the set $\{\mathbf{w}(t-r), r \geqslant 0\}$ is maximal in $\mathbf{S}$, the normal equations are

$$(\mathbf{h}(t) - \mathbf{v}(t), \mathbf{w}(t-r))_t = 0, \quad r \geqslant 0,$$

or

$$\int_{-\infty}^\infty [\mathbf{h}(t) - \mathbf{v}(t)]\,\mathbf{w}(t-r)\,dt = 0, \quad r \geqslant 0.$$

Because, for $r \geqslant 0$, $\mathbf{w}(t-r) = 0$ for all $t<0$, the normal equations become

$$\int_0^\infty [\mathbf{h}(t) - \mathbf{v}(t)]\, \mathbf{w}(t-r)\, dt = 0, \quad r \geqslant 0,$$

or

$$\int_0^\infty [\mathbf{h}(t)\, u(t) - \mathbf{v}(t)]\, \mathbf{w}(t-r)\, dt = 0, \quad r \geqslant 0,$$

or

$$\int_{-\infty}^\infty [\mathbf{h}(t)\, u(t) - \mathbf{v}(t)]\, \mathbf{w}(t-r)\, dt = 0, \quad r \geqslant 0. \tag{9.71}$$

Now $[\mathbf{h}(t)\, u(t) - \mathbf{v}(t)]$ is an element of wavelet space $\mathbf{L}_t^2(0,\infty)$. The normal equations (9.71) state that this element is orthogonal to $\mathbf{S}$. This orthogonality condition is necessary and sufficient that the mean-square error (9.70) is a minimum.

Proof. The mean-square error is

$$\begin{aligned} \|\mathbf{h}(t) - \mathbf{v}(t)\|_t^2 &= \|\boldsymbol{\varkappa}(t) * \mathbf{w}(t) - \boldsymbol{\gamma}(t) * \mathbf{w}(t)\|_t^2 \\ &= \|\mathbf{KW} - \boldsymbol{\Gamma}\mathbf{W}\|_\omega^2 = \|(\mathbf{K} - \boldsymbol{\Gamma})\, \mathbf{W}_0 \mathbf{A}\|_\omega^2 \\ &= \|(\mathbf{K} - \boldsymbol{\Gamma})\, \mathbf{W}_0\|_\omega^2 = \|\boldsymbol{\varkappa}(t) * \mathbf{w}_0(t) - \boldsymbol{\gamma}(t) * \mathbf{w}_0(t)\|_t^2 \\ &= \|\mathbf{h}_0(t) - \mathbf{v}_0(t)\|_t^2 \end{aligned}$$

where $\mathbf{h}_0 = \boldsymbol{\varkappa} * \mathbf{w}_0$, $\mathbf{v}_0 = \boldsymbol{\gamma} * \mathbf{w}_0$, and $\mathbf{w}_0$ is minimum-delay. Hence the $\boldsymbol{\gamma}$ for which $\boldsymbol{\gamma} * \mathbf{w}$ is the least-squares approximation to $\boldsymbol{\varkappa} * \mathbf{w}$ is exactly the same as the $\boldsymbol{\gamma}$ for which $\boldsymbol{\gamma} * \mathbf{w}_0$ is the least-squares approximation to $\boldsymbol{\varkappa} * \mathbf{w}_0$. But the wavelet $\boldsymbol{\gamma} * \mathbf{w}_0$ is an element of:

$$\text{Subspace spanned by} \quad \{\mathbf{w}_0(t-r), r \geqslant 0\} = \mathbf{L}_t^2(0,\infty).$$

Proceeding as before, we obtain the normal equations

$$\int_{-\infty}^\infty [\mathbf{h}_0(t)\, u(t) - \mathbf{v}_0(t)]\, \mathbf{w}_0(t-r)\, dt = 0, \quad r \geqslant 0,$$

or

$$[\mathbf{h}_0(t)\, u(t) - \mathbf{v}_0(t)] \perp \mathbf{w}_0(t-r), \quad r \geqslant 0.$$

Now $[\mathbf{h}_0(t)\, u(t) - \mathbf{v}_0(t)]$ is an element of $\mathbf{L}_t^2(0,\infty)$ and the set $\{\mathbf{w}_0(t-r), r \geqslant 0\}$ is maximal in $\mathbf{L}_t^2(0,\infty)$. Hence Theorem 8.20 applies, and consequently the normal equations are satisfied if and only if

$$[\mathbf{h}_0(t)\, u(t) - \mathbf{v}_0(t)] \equiv 0.$$

That is,

$$\boldsymbol{\gamma}(t) * \mathbf{w}_0(t) = [\boldsymbol{\varkappa}(t) * \mathbf{w}_0(t)]\, u(t).$$

Because $\mathbf{w}_0(t)$ is minimum-delay, Theorem 9.40(d) applies, and hence

$$\boldsymbol{\gamma}(t) = \{[\boldsymbol{\varkappa}(t) * \mathbf{w}_0(t)]\, u(t)\} * \mathbf{w}_0^{-1}(t)$$

is the least-squares realizable system. The fourier transform of $\boldsymbol{\gamma}(t)$ is

$$\boldsymbol{\Gamma}(i\omega) = \left\{\int_0^\infty dt\, \mathrm{e}^{-i\omega t}\, [\boldsymbol{\varkappa}(t) * \mathbf{w}_0(t)]\right\} \mathbf{W}_0^{-1}(i\omega).$$

Since

$$[\boldsymbol{\varkappa}(t) * \mathbf{w}_0(t)] = \frac{1}{2\pi}\int_{-\infty}^{\infty} d\mu\, \mathrm{e}^{i\mu t}\, \mathbf{K}(i\mu)\, \mathbf{W}_0(i\mu),$$

the gain-and-phase spectrum of the least-squares system is

$$\boldsymbol{\Gamma}(i\omega) = \left\{\int_0^\infty dt\, \mathrm{e}^{-i\omega t} \frac{1}{2\pi}\int_{-\infty}^{\infty} d\mu\, \mathrm{e}^{i\mu t}\, \mathbf{K}(i\mu)\, \mathbf{W}_0(i\mu)\right\} \mathbf{W}_0^{-1}(i\omega). \qquad \text{Q.E.D.}$$

Theorem 9.71. Let $\boldsymbol{\gamma}(t) * \mathbf{w}(t)$ be the least-squares approximation to $\boldsymbol{\varkappa}(t) * \mathbf{w}(t)$. Then the minimum mean-square error is

$$I_{\min} = \psi(\mathbf{w}) + \sigma(\mathbf{w})$$

where

$$\psi(\mathbf{w}) = \operatorname{tr}\int_{-\infty}^{0} |\boldsymbol{\varkappa}(t) * \mathbf{w}(t)|^2\, dt$$

and

$$\sigma(\mathbf{w}) = \operatorname{tr}\int_0^\infty |\boldsymbol{\varkappa}(t) * \mathbf{w}(t)|^2\, dt - \operatorname{tr}\int_0^\infty |\boldsymbol{\varkappa}(t) * \mathbf{w}_0(t)|^2\, dt$$

where $\mathbf{w}_0(t)$ is minimum-delay with the same square-gain as $\mathbf{w}(t)$.

Proof. By Theorem 8.31, we have

$$I_{\min} = \|\boldsymbol{\varkappa}(t) * \mathbf{w}(t)\|_t^2 - \|\boldsymbol{\gamma}(t) * \mathbf{w}(t)\|_t^2.$$

But

$$\begin{aligned}\|\boldsymbol{\gamma} * \mathbf{w}\|_t^2 &= \|\boldsymbol{\Gamma}\mathbf{W}\|_\omega^2 = \|\boldsymbol{\Gamma}\mathbf{W}_0\|_\omega^2 = \|\boldsymbol{\gamma} * \mathbf{w}_0\|_t^2 \\ &= \|[\boldsymbol{\varkappa} * \mathbf{w}_0]\, u(t) * \mathbf{w}_0^{-1} * \mathbf{w}_0\|_t^2 = \|[\boldsymbol{\varkappa} * \mathbf{w}_0]\, u(t)\|_t^2.\end{aligned}$$

Also

$$[\boldsymbol{\varkappa}(t) * \mathbf{w}(t)]\, u(-t) \perp [\boldsymbol{\varkappa}(t) * \mathbf{w}(t)]\, u(t).$$

Therefore

$$I_{\min} = \|[\boldsymbol{\varkappa} * \mathbf{w}]\, u(-t)\|_t^2 + \|[\boldsymbol{\varkappa} * \mathbf{w}]\, u(t)\|_t^2 - \|[\boldsymbol{\varkappa} * \mathbf{w}_0]\, u(t)\|_t^2.$$

We see that the first term is $\psi(\mathbf{w})$ and that the last two terms constitute $\sigma(\mathbf{w})$. Q.E.D.

We note that $[\boldsymbol{\varkappa} * \mathbf{w}]\, u(t)$ is in wavelet space $\mathbf{L}_t^2(0, \infty)$, and $\boldsymbol{\gamma} * \mathbf{w}$ is in $\mathbf{S} = \text{S.s.b.}\{\mathbf{w}(t-r), r \geqslant 0\}$.

Theorem 9.72. In Theorems 9.70 and 9.71:

(a) $[\boldsymbol{\varkappa} * \mathbf{w}]\, u(t)$ is the least-squares approximation to $\boldsymbol{\varkappa} * \mathbf{w}$, with minimum mean-square error $\psi(\mathbf{w})$.

(b) $\boldsymbol{\gamma} * \mathbf{w}$ is the least-squares approximation to $[\boldsymbol{\varkappa} * \mathbf{w}]\, u(t)$, with minimum mean-square error $\sigma(\mathbf{w})$.

Proof. By inspection, we see that the first-stage error

$$[\boldsymbol{\varkappa} * \mathbf{w}] - [\boldsymbol{\varkappa} * \mathbf{w}]\, u(t) = [\boldsymbol{\varkappa} * \mathbf{w}]\, u(-t)$$

is orthogonal to $\mathbf{L}_t^2(0, \infty)$, and thus condition (a) is established. Since $\boldsymbol{\gamma}(t)$ was chosen to minimize the mean-square error (9.70), the normal equations (9.71) are satisfied. That is, the second-stage error

$$\mathbf{h}(t)\, u(t) - \mathbf{v}(t) = [\boldsymbol{\varkappa} * \mathbf{w}]\, u(t) - \boldsymbol{\gamma} * \mathbf{w}$$

is orthogonal to $\mathbf{S}$, and thus condition (b) is established. Q.E.D.

Theorem 9.73. In Theorem 9.71, for all $\boldsymbol{\varkappa}(t)$, $\sigma(\mathbf{w}) \geqslant 0$, and $= 0$ if and only if $\mathbf{w}(t)$ is minimum-delay.

Proof. From Theorem 9.72, we have

$$\sigma(\mathbf{w}) = \| [\boldsymbol{\varkappa} * \mathbf{w}]\, u(t) - \boldsymbol{\gamma} * \mathbf{w} \|_t^2 \geqslant 0$$

where $\sigma(\mathbf{w})$ is the (minimum) mean-square error when the element $[\boldsymbol{\varkappa} * \mathbf{w}]\, u(t)$ in $\mathbf{L}_t^2(0, \infty)$ is approximated by the element $\boldsymbol{\gamma} * \mathbf{w}$ in $\mathbf{S}$. Hence $\sigma(\mathbf{w}) = 0$ for all $\boldsymbol{\varkappa}(t)$ if and only if the spaces $\mathbf{L}_t^2(0, \infty)$ and $\mathbf{S}$ coincide; i.e. if and only if $\mathbf{w}(t)$ is minimum-delay. Q.E.D.

Corollary 9.70. In Theorem 9.71, $I_{\min} = \psi(\mathbf{w}_0)$ where $\mathbf{w}_0$ is minimum-delay.

Theorem 9.74. Let $\mathbf{w}(t)$ be an invertible wavelet and let $\boldsymbol{\varkappa}(t)$ be any non-realizable system such that $\boldsymbol{\varkappa}(t) * \mathbf{w}(t)$ is L^2-stable. Then $\mathbf{w}(t)$ is minimum-delay if and only if

$$\psi(\mathbf{w}) = \operatorname{tr} \int_{-\infty}^{0} |\boldsymbol{\varkappa}(t) * \mathbf{w}(t)|^2\, dt$$

is a maximum.

Proof. From Theorems 9.71 and 9.73 and Corollary 9.70, we have

$$I_{\min} = \psi(\mathbf{w}_0) = \psi(\mathbf{w}) + \sigma(\mathbf{w}) \geqslant \psi(\mathbf{w})$$

with the equality holding if and only if $\sigma(\mathbf{w}) = 0$ for all $\boldsymbol{\varkappa}$; i.e. if and only if $\mathbf{w}$ is minimum-delay. Q.E.D.

9.8. The third minimum-delay wavelet theorem

We now have our third main theorem on minimum-delay.

Theorem 9.80 (the third minimum-delay wavelet theorem). Let $\mathbf{w}(t)$ be an invertible wavelet. Then each of the following conditions is necessary and sufficient that $\mathbf{w}(t)$ is minimum-delay.

(a) The partial energy

$$\operatorname{tr}\int_{-\infty}^{\tau} |\mathbf{w}(t)|^2\,dt = \operatorname{tr}\int_{0}^{\tau} |\mathbf{w}(t)|^2\,dt$$

of the wavelet $\mathbf{w}(t)$ is a *maximum* for all positive time $(\tau > 0)$.

(b) The partial energy

$$\operatorname{tr}\int_{-\infty}^{\tau} |\mathbf{w}^{-1}(t)|^2\,dt$$

of the inverse wavelet $\mathbf{w}^{-1}(t)$ is a *minimum* for all negative time $(\tau < 0)$, and this minimum is zero.

Proof. In the preceding section, let $\boldsymbol{\varkappa}(t) = \delta(t+\tau)\mathbf{I}$ where $\tau > 0$. Then $\psi(\mathbf{w})$ is the partial energy of $\mathbf{w}$, and condition (a) follows from Theorem 9.74. Condition (b) follows from Definition 9.20. Q.E.D.

The wavelet $w(t) = \sqrt{(2)}\,e^{-t}(1-2t)\,u(t)$ is non-minimum-delay and the wavelet $w_0(t) = \sqrt{(2)}\,e^{-t}u(t)$ is minimum-delay with the same gain (see Fig. 9.80). Let $\varkappa(t) = \delta(t+1)$. Then the least-squares realizable system of Theorem 9.70 is $\gamma(t) = e^{-1}\,\delta(t)$. Fig. 9.81 shows the least-squares error when $w(t)$ is the input; and Fig. 9.82, when $w_0(t)$ is the input. The least-squares errors shown in these two figures have the same (minimum) mean-square value.

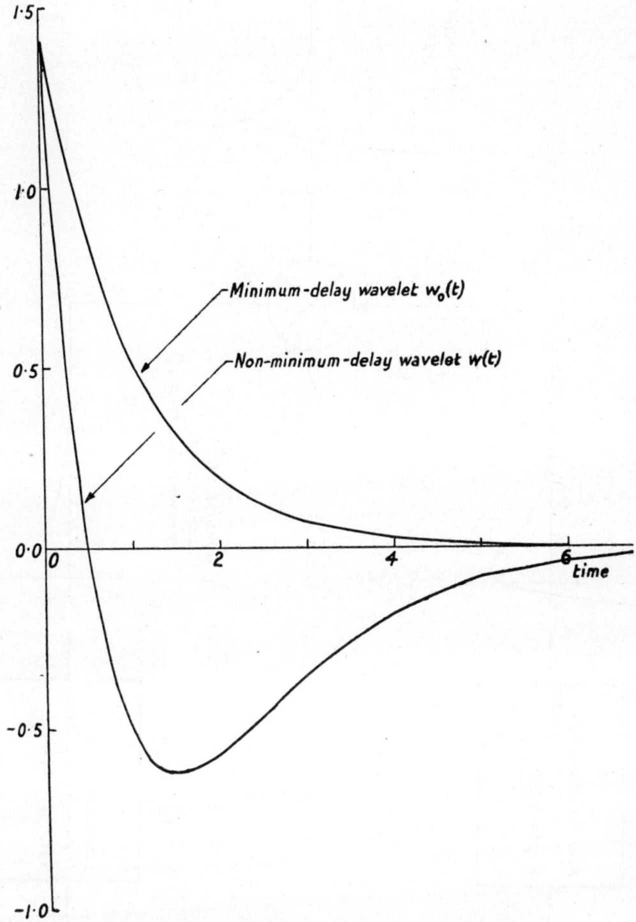

Fig. 9.80—Wavelets with the same gain

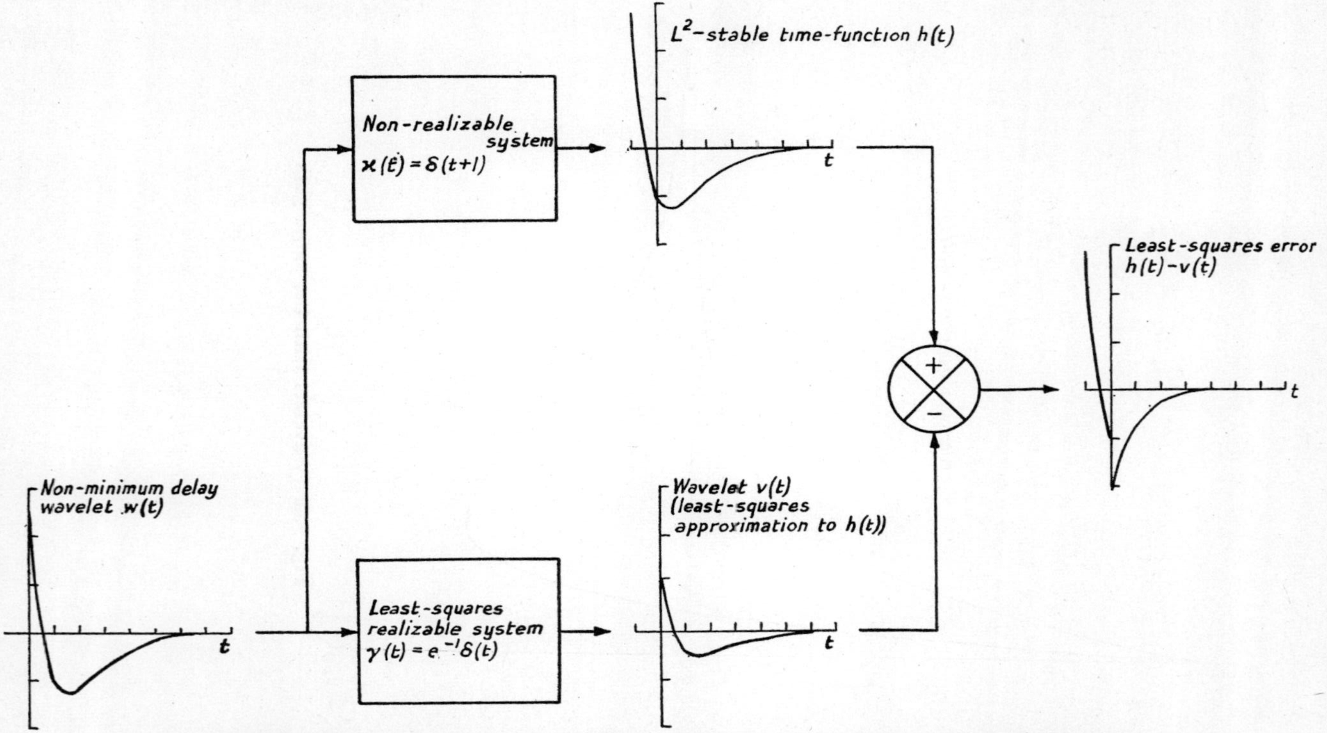

Fig. 9.81—Illustration of Theorem 9.70 with non-minimum delay wavelet as input

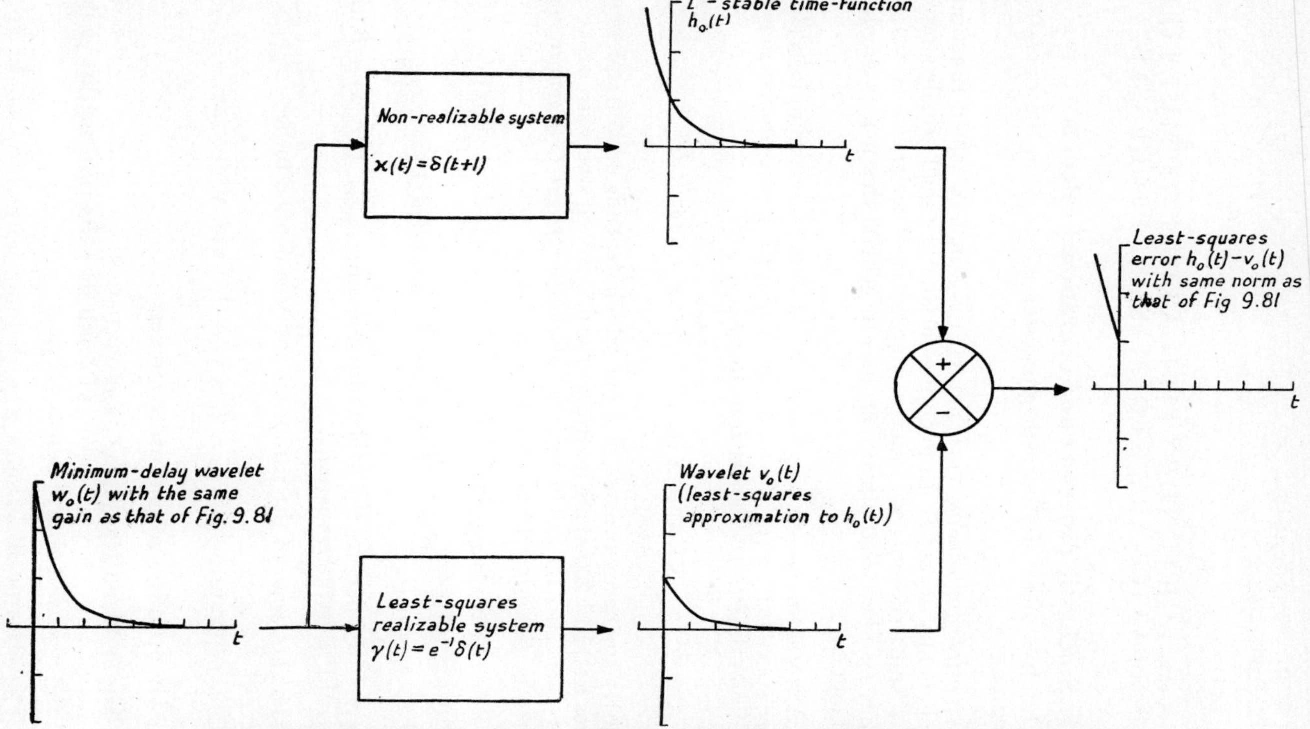

Fig. 9.82—Illustration of Theorem 9.70 with minimum-delay wavelet as input

CHAPTER X

STRUCTURAL IDENTIFICATION OF MULTIPLE STATIONARY STOCHASTIC PROCESSES

10.1. Stochastic processes and operational calculus

Our treatment of stochastic processes is based on the operational calculus.

A *variate* (or *random variable*) $\mathbf{x}(\xi)$ is a matrix-valued function of an elementary event ξ. In denoting variates, we shall usually suppress the ξ and write only $\mathbf{x}$. Since $\mathbf{x}$ is a matrix of scalar-valued variates, it may also be called a *multivariate*. In general this matrix would be rectangular.

A *stochastic process* is a family of variates. For example, $\mathbf{x}(t)$, or more specifically $\mathbf{x}(t, \xi)$, would represent a stochastic process depending upon time t. Its fourier transform

$$\mathbf{X}(i\omega) = \mathbf{X}(i\omega, \xi) = \int_{-\infty}^{\infty} \mathbf{x}(t, \xi)\, \mathrm{e}^{-i\omega t}\, dt$$

is also a stochastic process, and one which depends upon frequency ω. Since $\mathbf{x}(t)$ and $\mathbf{X}(i\omega)$ are matrices of scalar-valued stochastic processes, they may also be called *multiple stochastic processes*, or *multivariate stochastic processes*. In general these matrices would be rectangular.

The *expected value*, or *mean*, of a variate $\mathbf{x}$ is defined to be

$$E\{\mathbf{x}\} = E\{\mathbf{x}(\xi)\} = \int_{\text{all}\,\xi} \mathbf{x}(\xi)\, P(d\xi)$$

where $P(d\xi)$ denotes the probability measure of $d\xi$. If $E\{\mathbf{x}\} = \mathbf{0}$, then $\mathbf{x}$ is said to have *zero mean*.

The *Gramian* of the variates $\mathbf{x}$ and $\mathbf{y}$ is defined to be

$$((\mathbf{x}, \mathbf{y})) = (\mathbf{x}, \mathbf{y})_P = E\{\mathbf{x}\,\overline{\mathbf{y}}\} = \int_{\text{all}\,\xi} \mathbf{x}(\xi)\, \overline{\mathbf{y}(\xi)}\, P(d\xi).$$

The Gramian is a second-order moment.

The notation $((\mathbf{x}, \mathbf{y}))$ (i.e. double parentheses with no subscript) always denotes $(\mathbf{x}, \mathbf{y})_P$. Also $\|\mathbf{x}\|$ (with no subscript) in this chapter and the next will always denote $\|\mathbf{x}\|_P$.

The class of all variates $\mathbf{x}$ with finite norm form an $\mathbf{L}^2$ space, which we denote by $\mathbf{L}^2_P$. (See Section **8.5**.)

10.2. Multiple white noise

Definition 10.20. A stochastic process $\vartheta(t)$ with zero mean and autocovariance

$$((\vartheta(t), \vartheta(s))) = \delta(t-s)\mathbf{I} = \boldsymbol{\delta}(t-s)$$

is called (multiple) *white noise.*

Its fourier transform

$$\boldsymbol{\Theta}(i\omega) = \int_{-\infty}^{\infty} \vartheta(t)\,\mathrm{e}^{-i\omega t}\,dt$$

has zero mean and satisfies:

$$\begin{aligned}((\boldsymbol{\Theta}(i\omega), \boldsymbol{\Theta}(i\mu))) &= \left(\left(\int_{-\infty}^{\infty} \vartheta(t)\,\mathrm{e}^{-i\omega t}\,dt, \quad \int_{-\infty}^{\infty} \vartheta(s)\,\mathrm{e}^{-i\mu s}\,ds\right)\right)\\ &= \int_{-\infty}^{\infty} dt \int_{-\infty}^{\infty} ds\,\mathrm{e}^{-i\omega t}\,((\vartheta(t), \vartheta(s)))\,\mathrm{e}^{i\mu s}\\ &= \int_{-\infty}^{\infty} dt\,\mathrm{e}^{i(\mu-\omega)t}\,\mathbf{I} = 2\pi\delta(\mu-\omega)\mathbf{I} = 2\pi\boldsymbol{\delta}(\mu-\omega).\end{aligned}$$

The next to last equality follows from Theorem 2.53.

10.3. Multiple stationary stochastic processes

Definition 10.30. Let white noise $\vartheta(t)$ be the input to the system $\mathbf{b}(t)$. If $\mathbf{b}(t)$ is dissipative, then the output $\mathbf{x}(t)$ is called a *moving-average (multiple, second-order) stationary stochastic process.* If $\mathbf{b}(t) = \mathbf{w}(t)$ is realizable-dissipative-invertible (i.e. an invertible wavelet) then the output $\mathbf{x}(t)$ is called a *regular (multiple, second-order) stationary stochastic process.*

The diagram

$$\vartheta(t) \longrightarrow \boxed{\mathbf{b}(t)} \longrightarrow \mathbf{x}(t)$$

shows the *structure* of the stationary process $\mathbf{x}(t)$ (cf. Karhunen,1949).

We have

$$\mathbf{x}(t) = \mathbf{b}(t) * \vartheta(t) \leftrightarrow \mathbf{X}(i\omega) = \mathbf{B}(i\omega)\,\boldsymbol{\Theta}(i\omega).$$

That is

$$\mathbf{x}(t) = \int_{-\infty}^{\infty} \mathbf{b}(s)\,\vartheta(t-s)\,ds$$

which is called the *temporal representation* of $\mathbf{x}(t)$, and

$$\mathbf{x}(t) = \frac{1}{2\pi}\int_{-\infty}^{\infty} \mathrm{e}^{i\omega t}\,\mathbf{X}(i\omega)\,d\omega = \frac{1}{2\pi}\int_{-\infty}^{\infty} \mathrm{e}^{i\omega t}\,\mathbf{B}(i\omega)\,\boldsymbol{\Theta}(i\omega)\,d\omega$$

which is called the *spectral representation* (cf. Cramér, 1940, 1942) of $\mathbf{x}(t)$. Because $\boldsymbol{\vartheta}(t)$ and $\boldsymbol{\Theta}(i\omega)$ have zero mean, $\mathbf{x}(t)$ and $\mathbf{X}(i\omega)$ also have zero mean.

Fig. 10.30 shows in detail the structure when $\mathbf{x}(t)$ is a column vector with two scalar components $x_1(t)$ and $x_2(t)$, so that

$$x_1(t) = b_{11}(t) * \vartheta_1(t) + b_{12}(t) * \vartheta_2(t);$$
$$x_2(t) = b_{21}(t) * \vartheta_1(t) + b_{22}(t) * \vartheta_2(t).$$

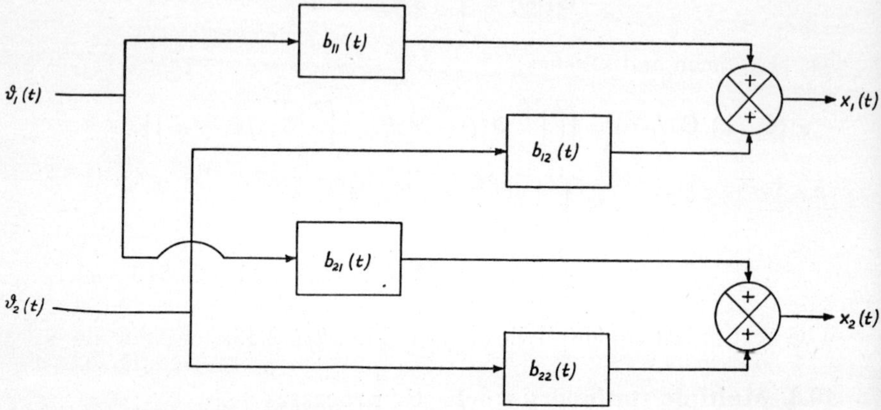

Fig. 10.30—Structure of a bivariate stationary stochastic process

The *autocovariance* of the stationary process $\mathbf{x}(t)$ is defined to be

$$\boldsymbol{\phi}(s) = E\{\mathbf{x}(t+s)\overline{\mathbf{x}(t)}\} = ((\mathbf{x}(t+s), \mathbf{x}(t))).$$

Theorem 10.30. The autocovariance of the stationary process $\mathbf{x}(t) = \mathbf{b}(t) * \boldsymbol{\vartheta}(t)$ is equal to the auto-product of the impulse response $\mathbf{b}(t)$; i.e.

$$\boldsymbol{\phi}(s) = ((\mathbf{x}(t+s), \mathbf{x}(t))) = (\mathbf{b}(t+s), \mathbf{b}(t))_t.$$

Proof. Using the temporal representation, we have

$$\begin{aligned}
\boldsymbol{\phi}(s) &= \left(\left(\int_{-\infty}^{\infty} \mathbf{b}(r)\boldsymbol{\vartheta}(t+s-r)\,dr, \quad \int_{-\infty}^{\infty} \mathbf{b}(\tau)\boldsymbol{\vartheta}(t-\tau)\,d\tau\right)\right) \\
&= \int_{-\infty}^{\infty} dr \int_{-\infty}^{\infty} d\tau\, \mathbf{b}(r)((\boldsymbol{\vartheta}(t+s-r), \boldsymbol{\vartheta}(t-\tau)))\overline{\mathbf{b}(\tau)} \\
&= \int_{-\infty}^{\infty} dr \int_{-\infty}^{\infty} d\tau\, \mathbf{b}(r)[\delta(s-r+\tau)\mathbf{I}]\overline{\mathbf{b}(\tau)} \\
&= \int_{-\infty}^{\infty} d\tau\, \mathbf{b}(\tau+s)\overline{\mathbf{b}(\tau)} = (\mathbf{b}(t+s), \mathbf{b}(t))_t. \qquad \text{Q.E.D.}
\end{aligned}$$

The above theorem shows that the white noise is averaged out (i.e. destroyed) in forming the autocovariance of the process.

If $\boldsymbol{\phi}(s)$ is the autocovariance of $\mathbf{x}(t)$, then the *spectral density* $\boldsymbol{\Phi}(i\omega)$ of $\mathbf{x}(t)$ is defined by

$$\boldsymbol{\phi}(s) = \frac{1}{2\pi}\int_{-\infty}^{\infty} e^{i\omega t}\,\boldsymbol{\Phi}(i\omega)\,d\omega \leftrightarrow \boldsymbol{\Phi}(i\omega).$$

Theorem 10.31. The spectral density of the stationary process $\mathbf{X}(i\omega) = \mathbf{B}(i\omega)\,\boldsymbol{\Theta}(i\omega)$ is equal to the square-gain of the system $\mathbf{B}(i\omega)$; i.e.

$$\boldsymbol{\Phi}(i\omega) = \mathbf{B}(i\omega)\,\overline{\mathbf{B}(i\omega)} = |\mathbf{B}(i\omega)|^2.$$

Proof. Since $\mathbf{b}(t+s) \leftrightarrow e^{i\omega s}\,\mathbf{B}(i\omega)$, we have

$$\boldsymbol{\phi}(s) = (\mathbf{b}(t+s), \mathbf{b}(t))_t = (e^{i\omega s}\,\mathbf{B}(i\omega), \mathbf{B}(i\omega))_\omega$$
$$= \frac{1}{2\pi}\int_{-\infty}^{\infty} e^{i\omega s}\,\mathbf{B}(i\omega)\,\overline{\mathbf{B}(i\omega)}\,d\omega. \qquad \text{Q.E.D.}$$

Using this theorem, we have

$$E\{\mathbf{X}(i\omega)\overline{\mathbf{X}(i\mu)}\} = E\{\mathbf{B}(i\omega)\,\boldsymbol{\Theta}(i\omega)\,\overline{\boldsymbol{\Theta}(i\mu)}\,\overline{\mathbf{B}(i\mu)}\}$$
$$= \mathbf{B}(i\omega)\,[2\pi\delta(\mu-\omega)\,\mathbf{I}]\,\overline{\mathbf{B}(i\mu)}$$
$$= 2\pi\delta(\mu-\omega)\,\mathbf{B}(i\omega)\,\overline{\mathbf{B}(i\mu)} = 2\pi\delta(\mu-\omega)\,\boldsymbol{\Phi}(i\omega)$$

(since $\delta(\mu-\omega) = 0$ except when $\mu = \omega$, i.e. when $\overline{\mathbf{B}(i\mu)} = \overline{\mathbf{B}(i\omega)}$).

10.4. Smoothing and stochastic integrals

White noise $\boldsymbol{\vartheta}(t)$, its fourier transform $\boldsymbol{\Theta}(i\omega)$, and $\mathbf{X}(i\omega)$ do not have finite norms, and so are not elements of $\mathbf{L}_P^2$. Nevertheless, they are essential to the stochastic operational calculus, where they usually enter not as end results, but in intermediate steps. In this respect, they play a role similar to that of the impulse function $\delta(t)$ in the non-stochastic operational calculus.

More specifically, processes such as $\boldsymbol{\vartheta}(t)$, $\boldsymbol{\Theta}(i\omega)$, and $\mathbf{X}(i\omega)$ usually appear inside integrals which *smooth* these processes, so that the integral (called a *stochastic integral*) has finite norm. For example, the regular stationary process

$$\mathbf{x}(t) = \int_0^{\infty} \mathbf{w}(s)\,\boldsymbol{\vartheta}(t-s)\,ds$$

has finite norm, and this integral (i.e. the temporal representation) is a stochastic integral. We may say that the white noise $\boldsymbol{\vartheta}(t)$ is smoothed by the system $\mathbf{w}(t)$ to yield a process $\mathbf{x}(t)$ with finite norm. Other examples of stochastic integrals are the spectral representation involving $\boldsymbol{\Theta}(i\omega)$ and the spectral representation involving $\mathbf{X}(i\omega)$.

Let $\delta_\epsilon(t)$ be the scalar-valued wavelet

$$\delta_\epsilon(t) = \frac{u(t)-u(t-\epsilon)}{\epsilon} = \begin{cases} 1/\epsilon & \text{for } 0 \leqslant t \leqslant \epsilon \\ 0 & \text{otherwise} \end{cases}$$

and let $\boldsymbol{\delta}_\epsilon(t) = \delta_\epsilon(t)\mathbf{I}$. We see that

$$\boldsymbol{\delta}_\epsilon(t) \to \boldsymbol{\delta}(t) \quad \text{as} \quad \epsilon \to 0.$$

For any function $\mathbf{f}(t)$ we shall use the notation

$$\langle \mathbf{f}(t) \rangle = \mathbf{f}(t) * \boldsymbol{\delta}_\epsilon(t) = \frac{1}{\epsilon}\int_0^\epsilon \mathbf{f}(t-r)\,dr = \frac{1}{\epsilon}\int_{t-\epsilon}^t \mathbf{f}(s)\,ds.$$

We see that $\langle \mathbf{f}(t) \rangle$ is a uniform smoothing of values of $\mathbf{f}(s)$ in the interval $t-\epsilon \leqslant s \leqslant t$, and we shall call $\langle \mathbf{f}(t) \rangle$ the *uniformly smoothed function.*

Because $\boldsymbol{\delta}_\epsilon(t)$ is an invertible wavelet, we have

Theorem 10.40. Uniformly smoothed white noise, $\langle \boldsymbol{\vartheta}(t) \rangle$, is a regular stationary process.

10.5. The random wavelet representation

Because an invertible wavelet $\mathbf{w}(t)$ is zero for $t<0$, we may consider the time instant $t=0$ as the *arrival time* of $\mathbf{w}(t)$. Similarly, we may consider the time instant $t=r$ as the arrival time of $\mathbf{w}(t-r)$. Let the strength, or amplification, of $\mathbf{w}(t-r)$ be white noise $\boldsymbol{\vartheta}(r)$. We shall then call $\mathbf{w}(t-r)\boldsymbol{\vartheta}(r)$ the *random wavelet* that arrives at time r. The strengths of random wavelets with different arrival times are uncorrelated, i.e. $E\{\boldsymbol{\vartheta}(r)\overline{\boldsymbol{\vartheta}(s)}\} = \mathbf{0}$ for $r \neq s$. Let $\mathbf{x}(t)$ be additively composed of all the random wavelets that have arrived up to time t; i.e.

$$\mathbf{x}(t) = \int_{-\infty}^t \mathbf{w}(t-r)\boldsymbol{\vartheta}(r)\,dr.$$

This equation is the temporal representation of a regular stationary process $\mathbf{x}(t)$. Hence for a regular stationary process, the temporal representation may also be called the *random wavelet representation.* In this representation, the wavelet $\mathbf{w}(t)$ represents the "dynamics" of the process, whereas the white noise $\boldsymbol{\vartheta}(t)$ represents the "random" nature of the process.

10.6. Ergodic theorems

In many applications it is possible to observe (over time) only one *fixed realization* of a stochastic process; i.e. to observe $\mathbf{x}(t, \xi_0)$ over time t, where the elementary event ξ_0 is fixed. The *time-average* of a fixed realization $\mathbf{x}(t, \xi_0)$ is defined to be

$$T\{\mathbf{x}(t, \xi_0)\} = \underset{\alpha \to -\infty}{\text{l.i.m.}} \frac{1}{t-\alpha}\int_\alpha^t \mathbf{x}(s, \xi_0)\,ds$$

(where l.i.m. is with reference to space $\mathbf{L}_P^2$). We shall let t represent present time, so that the time-average is an average of only past observations $\mathbf{x}(s, \xi_0)$, $s \leqslant t$.

The following *ergodic theorem* is the counterpart of Theorem 7.31 in Robinson (1959, p. 87), and the proof follows similar lines.

Theorem 10.60. If $\mathbf{x}(t)$ is a moving-average stationary process, then

$$T\{\mathbf{x}(t, \xi)\} = E\{\mathbf{x}(t, \xi)\} = 0$$

and

$$T\{\mathbf{x}(t, \xi)\, \mathrm{e}^{-i\omega t}\} = E\{\mathbf{x}(t, \xi)\, \mathrm{e}^{-i\omega t}\} = 0.$$

Theorem 10.61. If $\mathbf{x}(t)$ is a moving-average stationary process with autocovariance $\boldsymbol{\phi}(s)$, and if (for arbitrary s)

$$\mathbf{x}(t+s)\overline{\mathbf{x}(t)} - \boldsymbol{\phi}(s)$$

is a moving-average stationary process, then

$$\boldsymbol{\phi}(s) = E\{\mathbf{x}(t+s)\overline{\mathbf{x}(t)}\} = T\{\mathbf{x}(t+s)\overline{\mathbf{x}(t)}\}.$$

Proof. Applying Theorem 10.60 to $\mathbf{x}(t+s)\,\overline{\mathbf{x}(t)} - \boldsymbol{\phi}(s)$, we have

$$T\{\mathbf{x}(t+s)\overline{\mathbf{x}(t)} - \boldsymbol{\phi}(s)\} = 0$$

which gives

$$T\{\mathbf{x}(t+s)\overline{\mathbf{x}(t)}\} = T\{\boldsymbol{\phi}(s)\} = \boldsymbol{\phi}(s). \qquad \text{Q.E.D.}$$

If a process $\mathbf{x}(t)$ satisfies the hypothesis of Theorem 10.61, we shall say that the process is (second-order) *ergodic*. Thus, for an ergodic process, the autocovariance $\boldsymbol{\phi}(s)$ may be determined from the past values of one fixed realization.

10.7. The multiple stationary process identification problem

Definition 10.70. The *structure* of a regular stationary process $\mathbf{x}(t)$ is defined to be the components in the diagram

$$\boldsymbol{\vartheta}(t) \longrightarrow \boxed{\mathbf{w}(t)} \longrightarrow \mathbf{x}(t).$$

If $\mathbf{w}(t)$ is minimum-delay, then the structure is said to be a *minimum-delay structure.*

Suppose that the autocovariance $\boldsymbol{\phi}(s)$ (or the spectral density $\boldsymbol{\Phi}(i\omega)$) of a regular stationary process $\mathbf{x}(t)$ is known. Let $\mathbf{w}(t) \leftrightarrow \mathbf{W}(i\omega)$ be any invertible wavelet that satisfies

$$\int_0^\infty \mathbf{w}(t+s)\,\overline{\mathbf{w}(t)}\,dt = \boldsymbol{\phi}(s) \leftrightarrow \mathbf{W}(i\omega)\,\overline{\mathbf{W}(i\omega)} = \boldsymbol{\Phi}(i\omega),$$

and define the white noise $\boldsymbol{\vartheta}(t) \leftrightarrow \boldsymbol{\Theta}(i\omega)$ by

$$\boldsymbol{\vartheta}(t) = \mathbf{w}^{-1}(t) * \mathbf{x}(t) \leftrightarrow \boldsymbol{\Theta}(i\omega) = \mathbf{W}^{-1}(i\omega)\,\mathbf{X}(i\omega).$$

Then the structure

$$\vartheta(t) \longrightarrow \boxed{\text{System } \mathbf{w}(t)} \longrightarrow \mathbf{x}(t)$$

is an admissible structure for $\mathbf{x}(t)$. Hence we have

The multiple stationary process identification problem.

Time formulation: Which wavelet from the class of all invertible wavelets that satisfy

$$\mathbf{w}(t) \otimes \mathbf{w}(t) = \boldsymbol{\phi}(s)$$

should be chosen for the structure of $\mathbf{x}(t)$?

Frequency formulation: Which system from the class of all invertible, realizable, dissipative systems with square-gain $\boldsymbol{\Phi}(i\omega)$ should be chosen for the structure of $\mathbf{x}(t)$?

Our fourth main theorem on minimum-delay is

Theorem 10.70 (*the identification theorem*). A minimum-delay structure, and only a minimum-delay structure, can be determined from the autocovariance (or spectral density) of a regular, multiple stationary stochastic process.

Proof. The invertible wavelet $\mathbf{w}(t)$ has the canonical representation $\mathbf{w}(t) = \mathbf{w}_0(t) * \mathbf{a}(t)$. Hence all admissible structures have the form

$$\vartheta(t) \longrightarrow \boxed{\text{All-pass } \mathbf{a}(t)} - \vartheta_0(t) \rightarrow \boxed{\text{Minimum-delay } \mathbf{w}_0(t)} \longrightarrow \mathbf{x}(t).$$

(Because $\mathbf{a}(t)$ is all-pass, $\vartheta_0(t)$ is also white noise.) The minimum-delay wavelet $\mathbf{w}_0(t)$ can be determined (except for a trivial all-pass system) from the square-gain $\boldsymbol{\Phi}(i\omega)$. Letting $\mathbf{a}(t)$ be the unit matrix, we thereby obtain a minimum-delay structure. On the other hand, suppose that the structure is given and is not minimum-delay. Then $\mathbf{a}(t)$ is a non-trivial all-pass system, and it cannot be determined from $\boldsymbol{\Phi}(i\omega)$. Hence, no non-minimum-delay structure can be determined from the spectral density $\boldsymbol{\Phi}(i\omega)$. Q.E.D.

10.8. The first minimum-delay structure theorem

We now have our fifth main theorem on minimum-delay.

Theorem 10.80 (*the first minimum-delay structure theorem*). Let a regular multiple stationary process $\mathbf{x}(t)$ have the given structure

$$\vartheta(t) \longrightarrow \boxed{\mathbf{w}(t)} \longrightarrow \mathbf{x}(t).$$

Then each of the following conditions is necessary and sufficient that the structure is minimum-delay.

(a) The white noise $\vartheta(t)$ may be expressed in terms of the past values $\{\mathbf{x}(s), s \leqslant t\}$ and no future values $\{\mathbf{x}(s), t < s\}$; i.e.

$$\vartheta(t) = \int_{-\infty}^{t} \mathbf{w}^{-1}(t-s)\,\mathbf{x}(s)\,ds.$$

(b) $\mathbf{x}(t) \longrightarrow$ | Realizable system $\mathbf{w}^{-1}(t)$ | $\longrightarrow \vartheta(t)$.

(c) S.s.b. $\{\mathbf{x}(s), s \leqslant t\}$ = S.s.b. $\{\langle\vartheta(r)\rangle, \quad r \leqslant t\}$.

(Note: S.s.b. is an abbreviation of "subspace spanned by".)

Proof. Condition (a) is true if and only if $\mathbf{w}^{-1}(t)$ is realizable, i.e. if and only if $\mathbf{w}(t)$ is minimum-delay. Condition (b) is a restatement of condition (a). By Definition 10.30, we always have

$$\{\mathbf{x}(s), s \leqslant t\} \text{ is in S.s.b. } \{\langle\vartheta(r)\rangle, \quad r \leqslant t\}.$$

But

$$\{\langle\vartheta(r)\rangle, r \leqslant t\} \text{ is in S.s.b. } \{\mathbf{x}(s), \quad s \leqslant t\}$$

if and only if Condition (a) holds. Condition (c) thus follows.

Q.E.D.

CHAPTER XI

PREDICTING AND FILTERING SYSTEMS FOR MULTIPLE PROCESSES

11.1. Multiple multiple stationary processes

Two moving-average multiple stationary processes $\mathbf{x}_j(t)$ and $\mathbf{x}_k(t)$ are said to be *stationarily correlated* if their *cross-covariance*

$$\boldsymbol{\phi}_{jk}(s) = ((\mathbf{x}_j(t+s), \mathbf{x}_k(t)))$$

is independent of t. The moving-average multiple stationary processes $\mathbf{x}_1(t), \mathbf{x}_2(t), \ldots, \mathbf{x}_N(t)$ are said to constitute a *multiple multiple stationary process* if they are stationarily correlated in pairs.

Each covariance may be expressed as

$$\boldsymbol{\phi}_{jk}(s) = \frac{1}{2\pi}\int_{-\infty}^{\infty} \boldsymbol{\Phi}_{jk}(i\omega)\, e^{i\omega t}\, d\omega$$

where the $\boldsymbol{\Phi}_{jk}(i\omega)$ are *spectral densities.* For $j = k$, $\boldsymbol{\phi}_{jj}$ and $\boldsymbol{\Phi}_{jj}$ are the autocovariance and spectral density of $\mathbf{x}_j(t)$; whereas for $j \neq k$, $\boldsymbol{\phi}_{jk}$ and $\boldsymbol{\Phi}_{jk}$ are called the *cross-covariance* and *cross-spectral density* of $\mathbf{x}_j(t)$ and $\mathbf{x}_k(t)$. We have

$$\boldsymbol{\phi}_{jk}(s) = \overline{\boldsymbol{\phi}_{kj}(-s)} \leftrightarrow \boldsymbol{\Phi}_{jk}(i\omega) = \overline{\boldsymbol{\Phi}_{kj}(i\omega)}.$$

The matrix $[\boldsymbol{\Phi}_{jk}(i\omega)]$, $j, k, = 1, 2, \ldots, N$, called the *coherency matrix*, is a non-negative definite hermitian matrix that determines the spectral densities of all possible linear combinations of $\mathbf{x}_1(t), \ldots, \mathbf{x}_N(t)$.

11.2. Least-squares regression

In this chapter, we let:

$\mathbf{z}(t)$ and $\mathbf{x}(t)$ constitute a multiple multiple stationary process,
$\mathbf{x}(t) = \mathbf{w}(t) * \boldsymbol{\vartheta}(t)$ be regular,
$\boldsymbol{\lambda}(t)$ be the impulse response function of a system,
$\boldsymbol{\lambda}(t) * \mathbf{x}(t)$ be a moving-average multiple stationary process.

The processes $\mathbf{z}(t)$ and $\mathbf{x}(t)$ are in space $\mathbf{L}_P^2$. Let

$$\mathbf{S}_1 = \text{Subspace spanned by } \{\mathbf{x}(s), \quad -\infty < s < \infty\}$$

and

$$\mathbf{S}_2 = \text{Subspace spanned by } \{\mathbf{x}(s), \quad s \leqslant t\}.$$

The process $\boldsymbol{\lambda}(t) * \mathbf{x}(t)$ is in $\mathbf{S}_1$ if $\boldsymbol{\lambda}(t)$ is non-realizable, and is in $\mathbf{S}_2$ if $\boldsymbol{\lambda}(t)$ is realizable.

We shall call

$\mathbf{z}(t)$: the *desired output* or *regressand*,
$\mathbf{x}(t)$: the (*actual*) *input* or *regressor*,
$\boldsymbol{\lambda}(t)$: the *filtering system* or *regression-coefficient*,
$\boldsymbol{\lambda}(t) * \mathbf{x}(t)$: the (*actual*) *output* or *regression.*

We have

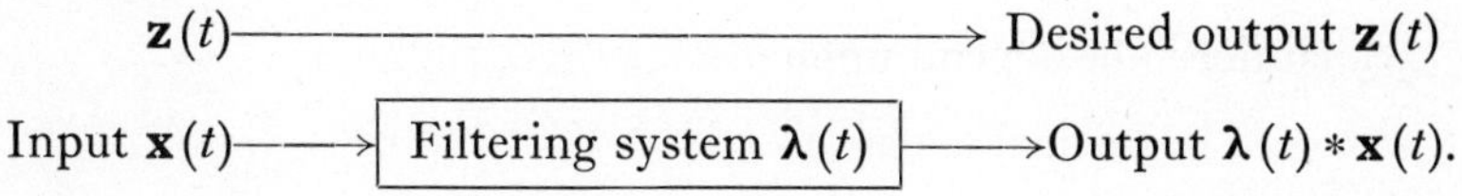

If the output $\boldsymbol{\lambda}(t) * \mathbf{x}(t)$ is the least-squares approximation to the desired output $\mathbf{z}(t)$; i.e. if the system $\boldsymbol{\lambda}(t)$ is determined so that the mean-square error

$$\|\mathbf{z}(t) - \boldsymbol{\lambda}(t) * \mathbf{x}(t)\|^2$$

is a minimum; then we call

$\boldsymbol{\lambda}(t)$: the *least-squares filtering system* or the *least-squares regression-coefficient*;
$\boldsymbol{\lambda}(t) * \mathbf{x}(t)$: the *least-squares output* or the *least-squares regression*;
$\mathbf{e}(t) = \mathbf{z}(t) - \boldsymbol{\lambda}(t) * \mathbf{x}(t)$: the *least-squares error* or the *least-squares residual*;

and we call the equation

$$\mathbf{z}(t) = \boldsymbol{\lambda}(t) * \mathbf{x}(t) + \mathbf{e}(t)$$

the *least-squares regression equation.*

11.3. Multiple filtering by a non-realizable system

We shall let $\mathbf{m}(t)$ be a moving-average, multiple, stationary stochastic process that is stationarily correlated with $\mathbf{x}(t)$. The process $\mathbf{m}(t)$ may be called the *message.*

In this section, we let:

the *desired output* be $\mathbf{m}(t+\alpha)$,
the *filtering system* be a non-realizable system $\boldsymbol{\varkappa}(t)$.

The real number α may be positive, zero, or negative. If $\alpha > 0$, the filtering system is called a *lead-filter*, or *predicting filter*; if $\alpha < 0$, the filtering system is called a *lag-filter.*

Theorem 11.30. The equation

$$\mathbf{m}(t+\alpha) = \boldsymbol{\varkappa}(t) * \mathbf{x}(t) + \mathbf{e}_1(t)$$

is the least-squares regression equation if and only if

$$\boldsymbol{\varkappa}(t) = \boldsymbol{\delta}(t+\alpha) * \boldsymbol{\phi}_{\mathbf{mx}}(t) * \boldsymbol{\phi}_{\mathbf{xx}}^{-1}(t),$$

$$\mathbf{K}(i\omega) = e^{i\omega\alpha}\, \boldsymbol{\Phi}_{\mathbf{mx}}(i\omega)\, \boldsymbol{\Phi}_{\mathbf{xx}}^{-1}(i\omega).$$

The minimum mean-square error is

$$\|\mathbf{e}_1(t)\|^2 = \operatorname{tr}\frac{1}{2\pi}\int_{-\infty}^{\infty} d\omega\, [\boldsymbol{\Phi}_{\mathbf{mm}} - \boldsymbol{\Phi}_{\mathbf{mx}}\boldsymbol{\Phi}_{\mathbf{xx}}^{-1}\boldsymbol{\Phi}_{\mathbf{xm}}]$$

which does not depend upon α.

Proof. The normal equations are

$$((\mathbf{m}(t+\alpha) - \boldsymbol{\varkappa}(t) * \mathbf{x}(t),\ \mathbf{x}(t-\tau))) = 0 \quad (-\infty < \tau < \infty),$$

which give:

$$((\mathbf{m}(t+\alpha), \mathbf{x}(t-\tau))) = \left(\left(\int_{-\infty}^{\infty} ds\, \boldsymbol{\varkappa}(s)\, \mathbf{x}(t-s), \mathbf{x}(t-\tau)\right)\right),$$

$$= \int_{-\infty}^{\infty} ds\, \boldsymbol{\varkappa}(s)\, ((\mathbf{x}(t-s), \mathbf{x}(t-\tau))),$$

or

$$\boldsymbol{\phi}_{\mathbf{mx}}(\tau+\alpha) = \int_{-\infty}^{\infty} ds\, \boldsymbol{\varkappa}(s)\, \boldsymbol{\phi}_{\mathbf{xx}}(\tau - s) = \boldsymbol{\varkappa}(\tau) * \boldsymbol{\phi}_{\mathbf{xx}}(\tau)$$

where $-\infty < \tau < \infty$. Hence the least-squares filtering system has impulse response

$$\boldsymbol{\varkappa}(\tau) = \boldsymbol{\phi}_{\mathbf{mx}}(\tau+\alpha) * \boldsymbol{\phi}_{\mathbf{xx}}^{-1}(\tau).$$

The fourier transform of $\boldsymbol{\varkappa}(\tau)$ gives $\mathbf{K}(i\omega)$.

By Theorem 8.31, the minimum mean-square error is

$$\|\mathbf{e}_1(t)\|^2 = \|\mathbf{m}(t+\alpha)\|^2 - \|\boldsymbol{\varkappa}(t) * \mathbf{x}(t)\|^2.$$

But

$$\|\boldsymbol{\varkappa} * \mathbf{x}\|^2 = \|\boldsymbol{\varkappa} * \mathbf{w} * \boldsymbol{\vartheta}\|^2 = \|\boldsymbol{\varkappa} * \mathbf{w}\|_t^2 = \|\mathbf{KW}\|_\omega^2.$$

Now

$$\begin{aligned}\mathbf{KW} &= e^{i\omega\alpha}\, \boldsymbol{\Phi}_{\mathbf{mx}}\boldsymbol{\Phi}_{\mathbf{xx}}^{-1}\, \mathbf{W} \\ &= e^{i\omega\alpha}\, \boldsymbol{\Phi}_{\mathbf{mx}}(\mathbf{W}\overline{\mathbf{W}})^{-1}\mathbf{W} \\ &= e^{i\omega\alpha}\, \boldsymbol{\Phi}_{\mathbf{mx}}\overline{\mathbf{W}}^{-1}\mathbf{W}^{-1}\mathbf{W} \\ &= e^{i\omega\alpha}\, \boldsymbol{\Phi}_{\mathbf{mx}}\overline{\mathbf{W}^{-1}} = e^{i\omega\alpha}\, \mathbf{G}\end{aligned}$$

where we define $\mathbf{g} \leftrightarrow \mathbf{G}$ by

$$\mathbf{g}(t) = \boldsymbol{\phi}_{\mathbf{mx}}(t) * \overline{\mathbf{w}^{-1}(-t)} \leftrightarrow \mathbf{G}(i\omega) = \boldsymbol{\Phi}_{\mathbf{mx}}(i\omega)\, \overline{\mathbf{W}^{-1}(i\omega)}.$$

Hence

$$\|\varkappa * \mathbf{x}\|^2 = \|e^{i\omega\alpha}\mathbf{G}\|_\omega^2 = \|\mathbf{g}(t+\alpha)\|_t^2 = \|\mathbf{G}\|_\omega^2 = \|\mathbf{\Phi}_{\mathbf{mx}}\overline{\mathbf{W}^{-1}}\|^2$$

$$= \operatorname{tr}\frac{1}{2\pi}\int_{-\infty}^{\infty}\mathbf{\Phi}_{\mathbf{mx}}\overline{\mathbf{W}^{-1}}\,\mathbf{W}^{-1}\overline{\mathbf{\Phi}_{\mathbf{mx}}}\,d\omega.$$

We therefore have

$$\|\mathbf{e}_1(t)\|^2 = \operatorname{tr}\boldsymbol{\phi}_{\mathbf{mm}}(0) - \|\mathbf{g}(t+\alpha)\|_t^2 \tag{11.30}$$

which is

$$\|\mathbf{e}_1(t)\|^2 = \operatorname{tr}\frac{1}{2\pi}\int_{-\infty}^{\infty}\mathbf{\Phi}_{\mathbf{mm}}(i\omega)\,d\omega - \operatorname{tr}\frac{1}{2\pi}\int_{-\infty}^{\infty}\mathbf{\Phi}_{\mathbf{mx}}(\mathbf{W}\overline{\mathbf{W}})^{-1}\mathbf{\Phi}_{\mathbf{xm}}\,d\omega.$$

Q.E.D.

11.4. Multiple filtering by a realizable system

In this section, we let:

the *desired output* be the moving-average stationary process given by $\varkappa(t) * \mathbf{x}(t)$ where $\varkappa(t)$ is a non-realizable system,

the *filtering system* be a realizable system $\boldsymbol{\gamma}(t)$.

Theorem 11.40. The equation

$$\varkappa(t) * \mathbf{x}(t) = \boldsymbol{\gamma}(t) * \mathbf{x}(t) + \mathbf{e}_2(t)$$

is the least-squares regression equation if and only if

$$\boldsymbol{\gamma}(t) = \{[\varkappa(t) * \mathbf{w}_0(t)]\,u(t)\} * \mathbf{w}_0^{-1}(t),$$

$$\mathbf{\Gamma}(i\omega) = \left\{\int_0^{\infty} e^{-i\omega t}[\varkappa(t) * \mathbf{w}_0(t)]\,dt\right\}\mathbf{W}_0^{-1}(i\omega)$$

where $\mathbf{w}_0(t) \leftrightarrow \mathbf{W}_0(i\omega)$ is minimum-delay with square-gain $\mathbf{\Phi}_{\mathbf{xx}}(i\omega)$. The minimum mean-square error is

$$\|\mathbf{e}_2(t)\|^2 = \operatorname{tr}\int_{-\infty}^{0}|\varkappa(t) * \mathbf{w}_0(t)|^2\,dt.$$

Proof. Let $\mathbf{x}(t)$ have the minimum-delay structure

$$\mathbf{x}(t) = \mathbf{w}_0(t) * \boldsymbol{\vartheta}_0(t).$$

Hence by Theorem 10.80,

$$\mathbf{S}_2 = \text{subspace spanned by } \{\langle \boldsymbol{\vartheta}_0(t-\tau)\rangle, \quad \tau \geqslant 0\}.$$

For brevity, let

$$\mathbf{f}_0 = \varkappa * \mathbf{w}_0 - \boldsymbol{\gamma} * \mathbf{w}_0.$$

Then $\mathbf{e}_2 = \mathbf{f}_0 * \boldsymbol{\vartheta}_0$, and hence the normal equations are

$$((\mathbf{e}_2(t), \langle \boldsymbol{\vartheta}_0(t-\tau)\rangle)) = ((\mathbf{f}_0(t) * \boldsymbol{\vartheta}_0(t), \langle \boldsymbol{\vartheta}_0(t-\tau)\rangle)) = 0 \text{ for } \tau \geqslant 0.$$

That is, for $\tau \geqslant 0$,

$$0 = \left(\left(\int_{-\infty}^{\infty} \mathbf{f}_0(s)\,\boldsymbol{\vartheta}_0(t-s)\,ds,\quad \frac{1}{\epsilon}\int_0^{\epsilon} \boldsymbol{\vartheta}_0(t-\tau-r)\,dr\right)\right)$$

$$= \frac{1}{\epsilon}\int_0^{\epsilon} dr \int_{-\infty}^{\infty} ds\,\mathbf{f}_0(s)\,((\boldsymbol{\vartheta}_0(t-s), \boldsymbol{\vartheta}_0(t-\tau-r)))$$

$$= \frac{1}{\epsilon}\int_0^{\epsilon} dr \int_{-\infty}^{\infty} ds\,\mathbf{f}_0(s)\,\delta(\tau+r-s)$$

$$= \frac{1}{\epsilon}\int_0^{\epsilon} dr\,\mathbf{f}_0(\tau+r) = \langle \mathbf{f}_0(\tau+\epsilon)\rangle.$$

Hence $\mathbf{f}_0(t) = 0$ for $t \geqslant 0$, or

$$\mathbf{f}_0(t)\,u(t) = [\boldsymbol{\varkappa}(t) * \mathbf{w}_0(t) - \boldsymbol{\gamma}(t) * \mathbf{w}_0(t)]\,u(t) = 0 \quad (-\infty < t < \infty),$$

which gives (since $[\boldsymbol{\gamma} * \mathbf{w}_0]\,u(t) = \boldsymbol{\gamma} * \mathbf{w}_0$)

$$\boldsymbol{\gamma}(t) * \mathbf{w}_0(t) = [\boldsymbol{\varkappa}(t) * \mathbf{w}_0(t)]\,u(t).$$

Since $\mathbf{w}_0(t)$ is minimum-delay, Theorem 9.40(d) applies. Thus the normal equations are satisfied if and only if

$$\boldsymbol{\gamma}(t) = \{[\boldsymbol{\varkappa}(t) * \mathbf{w}_0(t)]\,u(t)\} * \mathbf{w}_0^{-1}(t)$$

where $\mathbf{w}_0(t)$ is minimum-delay. This result is the same result as that of Theorem 9.70. In fact, the least-squares problem of this theorem is isomorphic to the one of that theorem.

The least-squares residual is

$$\mathbf{e}_2(t) = \mathbf{f}_0(t) * \boldsymbol{\vartheta}_0(t) = [\mathbf{f}_0(t)\,u(-t)] * \boldsymbol{\vartheta}_0(t)$$

since $\mathbf{f}_0(t)\,u(t) = 0$. But

$$\mathbf{f}_0(t)\,u(-t) = [\boldsymbol{\varkappa}(t) * \mathbf{w}_0(t) - \boldsymbol{\gamma}(t) * \mathbf{w}_0(t)]\,u(-t) = [\boldsymbol{\varkappa}(t) * \mathbf{w}_0(t)]\,u(-t)$$

since $\boldsymbol{\gamma} * \mathbf{w}_0$ is one-sided. Hence the minimum mean-square error is

$$\|\mathbf{e}_2(t)\|^2 = \|[\boldsymbol{\varkappa}(t) * \mathbf{w}_0(t)]\,u(-t)\|_t^2$$

$$= \operatorname{tr}\int_{-\infty}^{0} |\boldsymbol{\varkappa}(t) * \mathbf{w}_0(t)|^2\,dt$$

(cf. Corollary 9.70). Q.E.D.

11.5. The general multiple filtering system using two-stage least-squares

In this section, we let:

the *desired output* be $\mathbf{m}(t+\alpha)$,

the *filtering system* be a realizable system $\boldsymbol{\gamma}(t)$.

The real number α may be positive, zero, or negative. If $\alpha > 0$, the filtering system is called a *lead-filter*, or *predicting filter*. If $\alpha < 0$, the filtering system is called a *lag-filter* (cf. Wiener, 1942, Chapter 3).

Theorem 11.50. The equation

$$\mathbf{m}(t+\alpha) = \boldsymbol{\gamma}(t) * \mathbf{x}(t) + \mathbf{e}(t)$$

is the least-squares regression equation if and only if

$$\boldsymbol{\gamma}(t) = \int_0^\infty \mathbf{g}_0(r+\alpha)\, \mathbf{w}_0^{-1}(t-r)\, dr,$$

$$\boldsymbol{\Gamma}(i\omega) = \left[\int_0^\infty \mathrm{e}^{-i\omega t}\, \mathbf{g}_0(t+\alpha)\, dt\right] \mathbf{W}_0^{-1}(i\omega)$$

where $\mathbf{w}_0(t) \leftrightarrow \mathbf{W}_0(i\omega)$ is minimum-delay with square-gain $\boldsymbol{\Phi}_{\mathbf{xx}}(i\omega)$, and

$$\mathbf{g}_0(t) = \boldsymbol{\phi}_{\mathbf{mx}}(t) * \overline{\mathbf{w}_0^{-1}(-t)} \leftrightarrow \mathbf{G}_0(i\omega) = \boldsymbol{\Phi}_{\mathbf{mx}}(i\omega)\, \overline{\mathbf{W}_0^{-1}(i\omega)}.$$

The minimum mean-square error is

$$\|\mathbf{e}(t)\|^2 = \operatorname{tr} \boldsymbol{\phi}_{\mathbf{mm}}(0) - \operatorname{tr} \int_\alpha^\infty |\mathbf{g}_0(s)|^2\, ds$$

where the first term is independent of α and the second term depends upon α.

Proof. This proof is an application of two-stage least-squares (Theorem 8.40). Let:

$\boldsymbol{\varkappa}(t) * \mathbf{x}(t)$ be the least-squares approximation to $\mathbf{m}(t+\alpha)$,

$\boldsymbol{\gamma}(t) * \mathbf{x}(t)$ be the least-squares approximation to $\boldsymbol{\varkappa}(t) * \mathbf{x}(t)$.

Then

$\boldsymbol{\gamma}(t) * \mathbf{x}(t)$ is the least-squares approximation to $\mathbf{m}(t+\alpha)$.

By Theorem 11.30, the first-stage least-squares system is

$$\boldsymbol{\varkappa}(t) = \boldsymbol{\delta}(t+\alpha) * \boldsymbol{\phi}_{\mathbf{mx}}(t) * \boldsymbol{\phi}_{\mathbf{xx}}^{-1}(t).$$

We now have

$$\begin{aligned}\boldsymbol{\varkappa}(t) * \mathbf{w}_0(t) &= \boldsymbol{\delta}(t+\alpha) * \boldsymbol{\phi}_{\mathbf{mx}}(t) * \boldsymbol{\phi}_{\mathbf{xx}}^{-1}(t) * \mathbf{w}_0(t)\\ &= \boldsymbol{\delta}(t+\alpha) * \boldsymbol{\phi}_{\mathbf{mx}}(t) * \overline{\mathbf{w}_0^{-1}(-t)} * \mathbf{w}_0^{-1}(t) * \mathbf{w}_0(t)\end{aligned}$$

because

$$\boldsymbol{\phi}_{\mathbf{xx}}^{-1}(t) \leftrightarrow \boldsymbol{\Phi}_{\mathbf{xx}}^{-1} = (\mathbf{W}_0 \overline{\mathbf{W}}_0)^{-1} = \overline{\mathbf{W}}_0^{-1} \mathbf{W}_0^{-1} = \overline{\mathbf{W}_0^{-1}}\, \mathbf{W}_0^{-1}.$$

Thus

$$\boldsymbol{\varkappa}(t) * \mathbf{w}_0(t) = \boldsymbol{\delta}(t+\alpha) * \mathbf{g}_0(t) = \mathbf{g}_0(t+\alpha)$$

where $\mathbf{g}_0 \leftrightarrow \mathbf{G}_0$ is defined by

$$\mathbf{g}_0(t) = \boldsymbol{\phi}_{\mathbf{mx}}(t) * \overline{\mathbf{w}_0^{-1}(-t)} \leftrightarrow \mathbf{G}_0 = \boldsymbol{\Phi}_{\mathbf{mx}} \overline{\mathbf{W}_0^{-1}}.$$

By Theorem 11.40, the second-stage least-squares system is

$$\begin{aligned}\boldsymbol{\gamma}(t) &= \{[\boldsymbol{\varkappa}(t) * \mathbf{w}_0(t)]\, u(t)\} * \mathbf{w}_0^{-1}(t)\\ &= \{\mathbf{g}_0(t+\alpha)\, u(t)\} * \mathbf{w}_0^{-1}(t) = \int_0^\infty \mathbf{g}_0(r+\alpha)\, \mathbf{w}_0(t-r)\, dr\end{aligned}$$

which is also the overall least-squares system.

By Theorem 11.30 (see equation (11.30) in the proof), the first-stage minimum mean-square error is

$$\|\mathbf{e}_1(t)\|^2 = \operatorname{tr} \boldsymbol{\phi}_{\mathbf{mm}}(0) - \|\mathbf{g}(t+\alpha)\|_t^2.$$

But

$$\begin{aligned}\|\mathbf{g}(t+\alpha)\|_t^2 &= \|e^{i\omega\alpha}\, \mathbf{G}\|_\omega^2 = \|e^{i\omega\alpha}\, \boldsymbol{\Phi}_{\mathbf{mx}} \overline{\mathbf{W}^{-1}}\|_\omega^2\\ &= \|e^{i\omega\alpha}\, \boldsymbol{\Phi}_{\mathbf{mx}} \overline{\mathbf{W}_0^{-1}}\|^2 = \|e^{i\omega\alpha}\, \mathbf{G}_0\|_\omega^2 = \|\mathbf{g}_0(t+\alpha)\|_t^2.\end{aligned}$$

By Theorem 11.40, the second-stage minimum mean-square error is

$$\|\mathbf{e}_2(t)\|^2 = \|[\boldsymbol{\varkappa}(t) * \mathbf{w}_0(t)]\, u(-t)\|_t^2 = \|\mathbf{g}_0(t+\alpha)\, u(-t)\|_t^2.$$

Hence the minimum mean-square error is

$$\begin{aligned}\|\mathbf{e}(t)\|^2 &= \|\mathbf{e}_1(t)\|^2 + \|\mathbf{e}_2(t)\|^2\\ &= \operatorname{tr} \boldsymbol{\phi}_{\mathbf{mm}}(0) - \|\mathbf{g}_0(t+\alpha)\|_t^2 + \|\mathbf{g}_0(t+\alpha)\, u(-t)\|_t^2\\ &= \operatorname{tr} \boldsymbol{\phi}_{\mathbf{mm}}(0) - \|\mathbf{g}_0(t+\alpha)\, u(t)\|_t^2\\ &= \operatorname{tr} \boldsymbol{\phi}_{\mathbf{mm}}(0) - \operatorname{tr} \int_0^\infty |\mathbf{g}_0(t+\alpha)|^2\, dt.\end{aligned}$$

Q.E.D.

Example 11.50. Let $\mathbf{q}'(t) = d\mathbf{q}(t)/dt$ be a moving-average stationary process, stationarily correlated with $\mathbf{x}(t)$. We wish to find the realizable system $\boldsymbol{\gamma}(t)$ with input $\mathbf{x}(t)$ and output the least-squares approximation to $\mathbf{q}'(t+\alpha)$ (cf. Wiener, 1942, Chapter 5). To do so, let $\mathbf{q}'(t) = \mathbf{m}(t)$ in Theorem 11.50. Since

$$\boldsymbol{\phi}_{\mathbf{q}'\mathbf{x}} = \boldsymbol{\phi}'_{\mathbf{qx}} \leftrightarrow \boldsymbol{\Phi}_{\mathbf{q}'\mathbf{x}} = i\omega\, \boldsymbol{\Phi}_{\mathbf{qx}}$$

we obtain

$$\mathbf{g}_0(t) = \boldsymbol{\phi}'_{\mathbf{qx}}(t) * \overline{\mathbf{w}_0^{-1}(-t)} \leftrightarrow \mathbf{G}_0(i\omega) = i\omega\, \boldsymbol{\Phi}_{\mathbf{qx}} \overline{\mathbf{W}_0^{-1}}.$$

Upon substitution of $\mathbf{g}_0$ into the expression for $\boldsymbol{\gamma}$ in Theorem 11.50, we obtain the least-squares system.

11.6. The general multiple predicting system

In this section, we let:

the *desired output* be $\mathbf{x}(t+\alpha)$,

the *filtering system* be a realizable system $\boldsymbol{\pi}(t)$.

The real number α is now required to be positive. Hence the filtering system $\boldsymbol{\pi}(t)$ may be called a *predicting system for lead* α. The time interval $(t, t+\alpha)$ where t is the present time is called the *prediction time-span*. (Cf. Wiener, 1942, Chapter 2.)

Theorem 11.60. The equation

$$\mathbf{x}(t+\alpha) = \boldsymbol{\pi}(t) * \mathbf{x}(t) + \mathbf{e}(t)$$

is the least-squares regression equation if and only if

$$\boldsymbol{\pi}(t) = \int_0^\infty \mathbf{w}_0(r+\alpha)\, \mathbf{w}_0^{-1}(t-r)\, dr,$$

$$\boldsymbol{\Pi}(i\omega) = \left[\int_0^\infty e^{-i\omega t}\, \mathbf{w}_0(t+\alpha)\, dt\right] \mathbf{W}_0^{-1}(i\omega)$$

where $\mathbf{w}_0(t) \leftrightarrow \mathbf{W}_0(i\omega)$ is minimum-delay with square-gain $\boldsymbol{\Phi}_{\mathbf{xx}}(i\omega)$. The minimum mean-square (prediction) error is given by the partial energy

$$\|\mathbf{e}(t)\|^2 = \operatorname{tr} \int_0^\alpha |\mathbf{w}_0(s)|^2\, ds.$$

Proof. This theorem follows immediately from Theorem 11.50 by the substitution $\mathbf{m}(t+\alpha) = \mathbf{x}(t+\alpha)$. Then $\mathbf{G}_0$ becomes

$$\mathbf{G}_0 = \boldsymbol{\Phi}_{\mathbf{xx}} \overline{\mathbf{W}_0^{-1}} = \mathbf{W}_0 \overline{\mathbf{W}}_0 \overline{\mathbf{W}_0^{-1}} = \mathbf{W}_0$$

so $\mathbf{g}_0$ becomes $\mathbf{w}_0$, and thus we obtain the least-squares predicting system $\boldsymbol{\pi}(t)$. Also $\boldsymbol{\phi}_{\mathbf{mm}}$ becomes $\boldsymbol{\phi}_{\mathbf{xx}}$, and hence

$$\|\mathbf{e}(t)\|^2 = \operatorname{tr}\left[\boldsymbol{\phi}_{\mathbf{xx}}(0) - \int_\alpha^\infty |\mathbf{w}_0(s)|^2\, ds\right]$$

$$= \operatorname{tr}\left[\int_0^\infty |\mathbf{w}_0(s)|^2\, ds - \int_\alpha^\infty |\mathbf{w}_0(s)|^2\, ds\right]. \qquad \text{Q.E.D.}$$

In Theorem 11.60 the least-squares filtering system $\boldsymbol{\pi}(t)$ may also be called the *least-squares predicting system*. Similarly, the least-squares regression $\boldsymbol{\pi}(t) * \mathbf{x}(t)$ may also be called the *least-squares prediction*; and the least-squares error $\mathbf{e}(t)$, the *least-squares prediction error*. See Fig. 11.60.

11.7. The second minimum-delay structure theorem

Our sixth main theorem on minimum-delay shows the connection between the concepts of minimum-delay and least-squares.

Theorem 11.70 (*the second minimum-delay structure theorem*). Let a regular multiple stationary process $\mathbf{x}(t)$ have the given structure

$$\vartheta(t) \longrightarrow \boxed{\mathbf{w}(t)} \longrightarrow \mathbf{x}(t).$$

Then each of the following conditions is necessary and sufficient that the structure is minimum-delay.

(a) The equation

$$\mathbf{x}(t+\alpha) = \int_0^\infty \mathbf{w}(s+\alpha)\,\vartheta(t-s)\,ds + \int_{-\alpha}^0 \mathbf{w}(s+\alpha)\,\vartheta(t-s)\,ds$$

is the least-squares regression equation where the first term on the right-hand side is the least-squares prediction and the second term is the least-squares prediction error.

(b) The least-squares prediction error is a smoothing of $\vartheta(r)$ only over the prediction time-span $t < r \leqslant t+\alpha$, and not over past times $-\infty \leqslant r \leqslant t$.

Proof. (a) By Theorem 11.60, the least-squares prediction is

$$\boldsymbol{\pi}(t) * \mathbf{x}(t) = [\mathbf{w}_0(t+\alpha)\,u(t)] * \mathbf{w}_0^{-1}(t) * \mathbf{w}(t) * \vartheta(t).$$

But

$$\int_0^\infty \mathbf{w}(s+\alpha)\,\vartheta(t-s)\,ds = [\mathbf{w}(t+\alpha)\,u(t)] * \vartheta(t).$$

The above two relations are the same if and only if $\mathbf{w}(t) = \mathbf{w}_0(t)\,\mathbf{A}_0$, where $\mathbf{A}_0$ is a trivial all-pass system, i.e. if and only if $\mathbf{w}(t)$ is minimum-delay. Q.E.D. (a)

(b) By Theorem 11.60, the least-squares prediction error is

$$\begin{aligned}
\mathbf{e}(t) &= \mathbf{x}(t+\alpha) - \boldsymbol{\pi}(t) * \mathbf{x}(t) \\
&= \int_{-\alpha}^\infty \mathbf{w}(s+\alpha)\,\vartheta(t-s)\,ds - \int_0^\infty [\boldsymbol{\pi}(s) * \mathbf{w}(s)]\,\vartheta(t-s)\,ds \\
&= \int_{-\alpha}^0 \mathbf{w}(s+\alpha)\,\vartheta(t-s)\,ds + \int_0^\infty \{\mathbf{w}(s+\alpha) - [\boldsymbol{\pi}(s) * \mathbf{w}(s)]\}\,\vartheta(t-s)\,ds.
\end{aligned}$$

The first term of this last expression is a smoothing of $\vartheta(t-s)$ over the interval $-\alpha \leqslant s < 0$; i.e. a smoothing of $\vartheta(r)$ over the prediction time-span $t < r \leqslant t+\alpha$. The second term is a smoothing of $\vartheta(t-s)$ over $0 \leqslant s < \infty$; i.e. a smoothing of $\vartheta(r)$ over the past times $-\infty < r \leqslant 0$. This second term vanishes if and only if

$$\mathbf{w}(s+\alpha)\,u(s) = \boldsymbol{\pi}(s) * \mathbf{w}(s) = [\mathbf{w}_0(s+\alpha)\,u(s)] * \mathbf{w}_0^{-1}(s) * \mathbf{w}(s);$$

i.e. if and only if $\mathbf{w}(s)$ is minimum-delay (cf. Condition (a) above).

Q.E.D. (b)

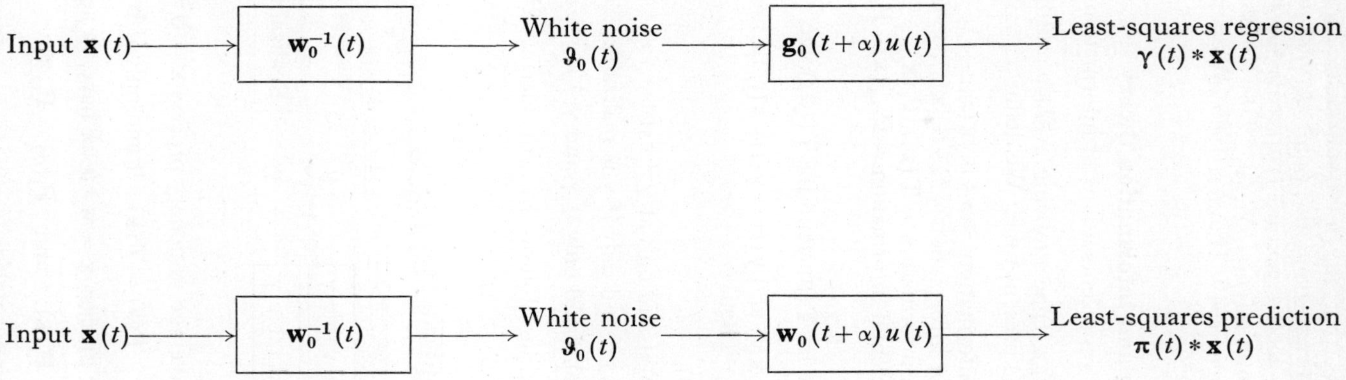

Fig. 11.60—Illustration of Theorems 11.50 and 11.60

This theorem shows that the minimum-delay structure is the structure that is appropriate from the point of view of prediction. That is, the minimum-delay structure represents the predictive (or Woldian) decomposition of $\mathbf{x}(t)$ (cf. Wold, 1938; Kolmogorov, 1941; Zasuhin, 1941; Karhunen, 1949; Hanner, 1949; Wiener and Masani, 1957, 1958).

11.8. Minimum-delay and information theory

A stochastic process represents a flow of information with time.

Definition 11.80. Let $\mathbf{x}(t)$ be a moving-average multiple stationary process. Let t be the present time. We define

$$\begin{aligned}
\mathbf{T}(\mathbf{x}) &= \text{total information} = \text{S.s.b.}\{\mathbf{x}(s), \quad -\infty < s < \infty\},\\
\mathbf{P}_t(\mathbf{x}) &= \text{present information} = \text{S.s.b.}\{\mathbf{x}(s), \quad s \leqslant t\},\\
\mathbf{F}_t(\mathbf{x}) &= \text{future information} = \mathbf{T}(\mathbf{x}) - \mathbf{P}_t(\mathbf{x}),\\
\mathbf{I}_{t,t+\alpha}(\mathbf{x}) &= \text{incremental information} = \mathbf{P}_{t+\alpha}(\mathbf{x}) - \mathbf{P}_t(\mathbf{x}),
\end{aligned}$$

(where $\alpha > 0$).

For white noise $\vartheta(t)$, we let $\mathbf{T}(\vartheta)$ stand for $\mathbf{T}(\langle\vartheta\rangle)$, etc.

Theorem 11.80. Let $\mathbf{x}(t)$ and $\mathbf{y}(t) = \gamma(t) * \mathbf{x}(t)$ be moving-average stationary processes. Then

(a) $\mathbf{T}(\mathbf{x}) \supset \mathbf{T}(\mathbf{y})$,
(b) $\mathbf{T}(\mathbf{x}) = \mathbf{T}(\mathbf{y})$ if and only if γ^{-1} exists,
(c) $\mathbf{P}_t(\mathbf{x}) \supset \mathbf{P}_t(\mathbf{y})$ if and only if γ is realizable,
(d) $\mathbf{P}_t(\mathbf{x}) = \mathbf{P}_t(\mathbf{y})$ if and only if γ and γ^{-1} are realizable.

Proof. The conditions follow by considering the following diagrams:

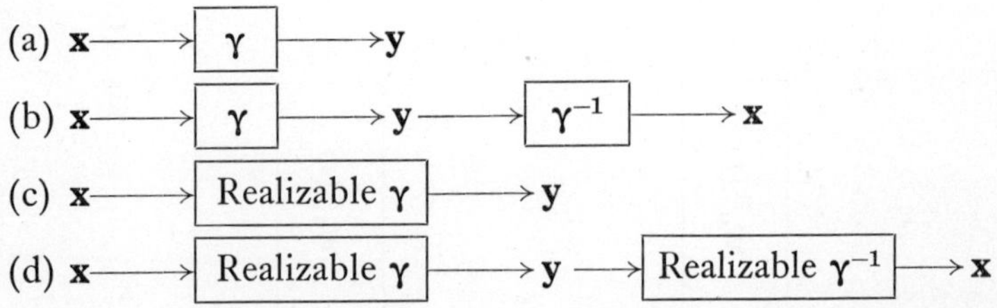

Q.E.D.

Hence for the moving-average stationary process $\mathbf{x} = \mathbf{b} * \vartheta$ we have

$$\mathbf{T}(\vartheta) \supset \mathbf{T}(\mathbf{x}); \quad \text{and} \quad \mathbf{T}(\vartheta) = \mathbf{T}(\mathbf{x}) \quad \text{if and only if } \mathbf{b}^{-1} \text{ exists.}$$

For a regular stationary process $\mathbf{x} = \mathbf{w} * \vartheta$ we have, in addition,

$$\mathbf{P}_t(\vartheta) \supset \mathbf{P}_t(\mathbf{x}); \quad \text{and} \quad \mathbf{P}_t(\vartheta) = \mathbf{P}_t(\mathbf{x})$$

if and only if $\mathbf{w}$ is minimum-delay.

Let $\mathbf{x}(t)$ be a moving-average stationary process, and let

$$\mathbf{y}(t) = \mathbf{x}(t) * \mathbf{a}(t)$$

where $\mathbf{a}(t)$ is an all-pass system. Then

$\mathbf{T}(\mathbf{x}) = \mathbf{T}(\mathbf{y})$, $\mathbf{P}_t(\mathbf{x}) \supset \mathbf{P}_t(\mathbf{y})$; and

$\mathbf{P}_t(\mathbf{x}) = \mathbf{P}_t(\mathbf{y})$ if and only if $\mathbf{a}(t)$ is trivial.

Theorem 11.81. Let $\mathbf{x}(t) = \mathbf{w}(t) * \vartheta(t)$ be a regular stationary process. Let $\mathbf{w}(t) = \mathbf{w}_0(t) * \mathbf{a}(t)$ where $\mathbf{w}_0(t)$ is minimum-delay and $\mathbf{a}(t)$ is all-pass. Then $\mathbf{P}_t(\mathbf{a} * \vartheta) = \mathbf{P}_t(\mathbf{x})$.

Proof. The proof follows from the diagram:

$$\mathbf{a} * \vartheta \longrightarrow \boxed{\text{Minimum-delay } \mathbf{w}_0} \longrightarrow \mathbf{x} \longrightarrow \boxed{\text{Realizable } \mathbf{w}_0^{-1}} \longrightarrow \mathbf{a} * \vartheta.$$

Corollary 11.80. If $\mathbf{x} = \mathbf{w} * \vartheta$ and $\mathbf{y} = \mathbf{v} * \vartheta$ where $\mathbf{w} = \mathbf{w}_0 * \mathbf{a}$ and $\mathbf{v} = \mathbf{v}_0 * \mathbf{a}$, then $\mathbf{P}_t(\mathbf{x}) = \mathbf{P}_t(\mathbf{y})$.

11.9. The third minimum-delay structure theorem

If $\mathbf{T}(\mathbf{x}) = \mathbf{T}(\gamma * \mathbf{x})$, we shall say that the system $\gamma(t)$ does not destroy the information contained in $\mathbf{x}(t)$. If $\mathbf{P}_t(\mathbf{x}) = \mathbf{P}_t(\gamma * \mathbf{x})$ we shall say that the system $\gamma(t)$ does not delay the information contained in $\mathbf{x}(t)$. The results of the preceding section may now be summarized in our seventh main theorem, which shows the connection between minimum-delay and information.

Theorem 11.90 (the third minimum-delay structure theorem). Let a regular multiple stationary process $\mathbf{x}(t)$ have the given structure

$$\vartheta(t) \longrightarrow \boxed{\mathbf{w}(t)} \longrightarrow \mathbf{x}(t).$$

Then each of the following conditions is necessary and sufficient that the structure is minimum-delay.

(a) $\mathbf{P}_t(\vartheta) = \mathbf{P}_t(\mathbf{x})$.

(b) $\mathbf{I}_{t,t+\alpha}(\vartheta) = \mathbf{I}_{t,t+\alpha}(\mathbf{x})$.

(c) In the transmission of information from input $\vartheta(t)$ to output $\mathbf{x}(t)$, the system $\mathbf{w}(t)$ neither destroys nor delays the information.

CHAPTER XII

NON-LINEAR FEEDBACK SYSTEMS

12.1. Feedback

Feedback occurs in, or may be introduced into, cybernetic systems to attain certain operating characteristics, to overcome limitations of physical components, or for other reasons. The *order* of a cybernetic system is a measure of the number of independent feedback loops. The higher the order, the more complex is the system.

The simplest feedback system is a system with a single closed loop. Let the input to the system be $x(t)$, let the impulse response of the system be $b(t)$ (called the closed-loop impulse response), and let the output of the system be $y(t)$:

$$x(t) \longrightarrow \boxed{\text{Closed-loop system } b(t)} \longrightarrow y(t).$$

The closed-loop system $b(t)$ internally consists of three components: the *feedback loop*, the *mixer*, and the *motor*. The term "motor" comes from the theory of servomechanisms, where this component is indeed a motor.

The *feedback loop* is a system whose input is $y(t)$; i.e. the output $y(t)$ of the closed loop system is the input of the feedback loop. Designate the impulse response of the feedback loop by $h(t)$, and its output by $z(t)$. Then we have:

$$z(t) \longleftarrow \boxed{\text{Feedback loop } h(t)} \longleftarrow y(t).$$

The *mixer* is a system with two inputs:

(1) $x(t)$, the input of the closed-loop system, and
(2) $z(t)$, the output of the feedback loop.

The output $e(t)$ of the mixer is the difference of its two inputs:

$$e(t) = x(t) - z(t).$$

The mixer output $e(t)$ is called the *actuating signal*. Schematically, we have:

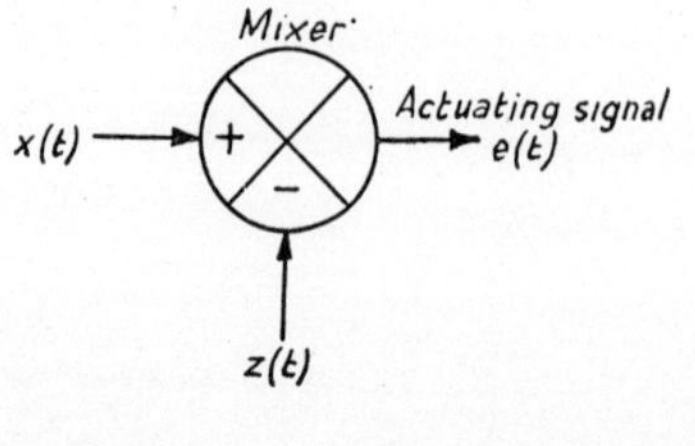

The *motor* is a system whose input is the actuating signal $e(t)$ and whose output is the output $y(t)$ of the closed-loop system. Let the impulse response of the motor be $g(t)$. Then we have:

$$e(t) \longrightarrow \boxed{\text{Motor } g(t)} \longrightarrow x(t).$$

Combining the diagrams for the feedback loop, mixer, and motor, we have the closed-loop system:

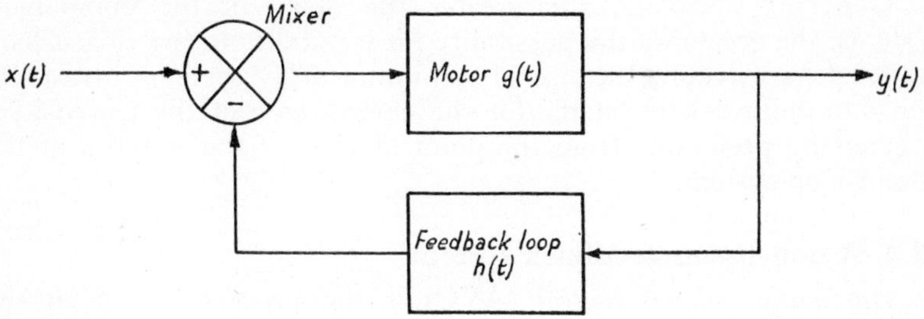

We have:

$$\begin{aligned}
&\text{Closed-loop system:} & y(t) &= x(t) * b(t)\\
&\text{Feedback loop:} & z(t) &= y(t) * h(t)\\
&\text{Mixer:} & e(t) &= x(t) - z(t)\\
&\text{Motor:} & y(t) &= e(t) * g(t)
\end{aligned}$$

or, in terms of laplace transforms:

$$\begin{aligned}
&\text{Closed-loop system:} & Y &= XB\\
&\text{Feedback loop:} & Z &= YH\\
&\text{Mixer:} & E &= X - Z\\
&\text{Motor:} & Y &= EG.
\end{aligned}$$

We now want to solve for the closed-loop transfer function B in terms of the components G and H. We have

$$Y = EG = (X - Z)G = (X - YH)G,$$

and, using $Y = XB$, we obtain

$$XB = (X - XBH)G,$$

which gives

$$B = (1 - BH)G,$$

or

$$B + BHG = G.$$

Therefore, the closed-loop transfer function $B(p)$ in terms of the motor transfer function $G(p)$ and the feedback loop transfer function $H(p)$ is

$$B(p) = \frac{G(p)}{1 + H(p)G(p)}.$$

For the closed-loop system $B(p)$ to be realizable and dissipative, all the poles of $B(p)$ must lie in the left half-plane (see Example 3.40). In other words, all the zeros of $1+H(p)G(p)$ must lie in the left half-plane, or equivalently all the -1 points of $H(p)G(p)$ must lie in the left half-plane.

A realizable system $B(p)$ whose impulse response $b(t)$ grows in magnitude without limit as time increases is said to be *unstable*. Such an unstable system has been exhibited in Example 3.42.

Generally speaking, the greater the delay in the component systems, the greater is the possibility for instability in the closed-loop system. As a result, the use of minimum-delay systems instead of non-minimum-delay systems for the components of the closed-loop is generally preferable from the point of view of the stability of the closed-loop system.

12.2. A non-linear feedback system

Automatic volume control (AVC) is the process by which the amplification of the input signal is controlled by the output, so that only small changes in the output result from large variations in the input. The control should depend upon the output amplitude level in such a way that it is fairly independent of the actual form of the signal. The speed of control should be fast enough to follow normal energy arrival, but not so fast as completely to distort the wavelet shapes.

A typical AVC system is a non-linear system such as the one shown in the following diagram:

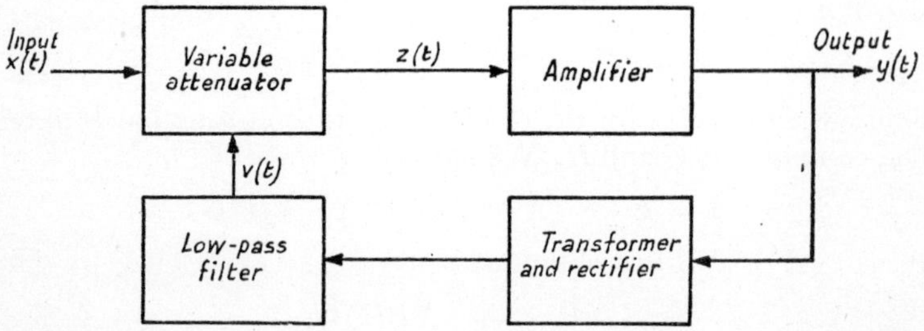

We shall assume that all time-functions, e.g. $x(t)$, $z(t)$, $y(t)$, and $v(t)$, are real.

The *variable attenuator* is a non-linear system with two inputs:

(1) the AVC system input $x(t)$, and
(2) the bias voltage $v(t)$ from the feedback loop.

The output $z(t)$ of the variable attenuator is given by

$$z(t) = Cx(t)\exp[-cv(t)]$$

where C and c are constants.

The *amplifier* may be assumed to have a constant gain A over the frequency band of interest, so its input $z(t)$ and its output $y(t)$ are related by

$$y(t) = Az(t).$$

The output $y(t)$ of the amplifier is the output of the AVC system.

The AVC system output $y(t)$ is fed-back into the *transformer-and-rectifier*. The transformer-and-rectifier is a non-linear system that produces full-wave rectification. In other words, if the input to the transformer-and-rectifier is $y(t)$, then its output is the absolute value $|y(t)|$.

Following the transformer-and-rectifier in the feedback loop is the *low-pass filter*, with impulse response function

$$h(t) = \kappa\, e^{-\kappa t} u(t)$$

where κ is a constant. The input of the low-pass filter is $|y(t)|$, and so its output is

$$v(t) = |y(t)| * h(t) = \int_{-\infty}^{t} \kappa\, e^{-\kappa(t-s)} |y(s)|\, ds.$$

Combining the above equations, we obtain the equation

$$y(t) = ACx(t) \exp\left[-c\int_{-\infty}^{t} \kappa\, e^{-\kappa(t-s)} |y(s)|\, ds\right]$$

that relates input $x(t)$ and output $y(t)$ of the closed-loop AVC system. This equation was given by Smith (1959) who considered the following envelope analysis.

Let the input $x(t)$ be the amplitude modulated sine wave

$$x(t) = f(t) \sin \omega_0 t$$

where ω_0 is a constant and $f(t) \geqslant 0$ is the envelope of the input. Then the output $y(t)$ will be approximately an amplitude-modulated sine wave of the same frequency ω_0; i.e.

$$y(t) \doteq g(t) \sin \omega_0 t$$

where $g(t) \geqslant 0$ is the envelope of the output. The frequency ω_0 is chosen to be well above the cut-off frequency of the low-pass filter. Consequently the sinusoidal component of frequency ω_0 may be assumed to be completely filtered out by the low-pass filter, and hence the convolution integral

$$v(t) = \int_{-\infty}^{t} \kappa\, e^{-\kappa(t-s)} |y(s)|\, ds = \int_{-\infty}^{t} \kappa\, e^{-\kappa(t-s)} g(s) |\sin \omega_0 s|\, ds$$

becomes

$$v(t) = \int_{-\infty}^{t} \kappa\, e^{-\kappa(t-s)} g(s)\, ds.$$

The Smith equation therefore becomes

$$g(t)\sin\omega_0 t = ACf(t)\cos\omega_0 t \exp\left[-c\int_{-\infty}^{t} \kappa\, e^{-\kappa(t-s)} g(s)\, ds\right]$$

which is

$$g(t) = ACf(t)\exp\left[-c\int_{-\infty}^{t} \kappa\, e^{-\kappa(t-s)} g(s)\, ds\right].$$

Let the constants f_0 and g_0 be the steady-state values of the input and output envelopes respectively; that is, let

$$g_0 = ACf_0 \exp\left[-c\int_{-\infty}^{t} \kappa\, e^{-\kappa(t-s)} g_0\, ds\right].$$

Taking the quotient of the above two equations, we have

$$\frac{g(t)}{g_0} = \frac{f(t)}{f_0}\exp\left\{-c\int_{-\infty}^{t} \kappa\, e^{-\kappa(t-s)} [g(s)-g_0]\, ds\right\}.$$

When x is close to 1, the approximation

$$\log x \doteq x-1$$

is good. For an AVC system, the variations of the output-envelope $g(t)$ compared to the steady-state value g_0 should be small. Hence $g(t)/g_0$ would be close to 1, yielding the approximation

$$\log\frac{g(t)}{g_0} \doteq \frac{g(t)}{g_0} - 1 = \frac{1}{g_0}[g(t)-g_0].$$

Hence we have

$$\frac{g(t)}{g_0} = \frac{f(t)}{f_0}\exp\left\{-c\int_{-\infty}^{t} \kappa\, e^{-\kappa(t-s)} g_0 \log\frac{g(t)}{g_0}\, ds\right\}.$$

Taking logarithms, we have

$$\log\frac{g(t)}{g_0} = \log\frac{f(t)}{f_0} - cg_0\int_{-\infty}^{t} \kappa\, e^{-\kappa(t-s)} \log\frac{g(s)}{g_0}\, ds.$$

If we define

$$\psi(t) = \log\frac{f(t)}{f_0}, \quad \gamma(t) = \log\frac{g(t)}{g_0}, \quad k = cg_0,$$

then we may write

$$\gamma(t) = \psi(t) - k\int_{-\infty}^{t} \kappa\, e^{-\kappa(t-s)} \gamma(s)\, ds,$$

which gives

$$\psi(t) = \gamma(t) + k[\kappa\, e^{-\kappa t} u(t)] * \gamma(t).$$

This equation is a linear equation that allows us to calculate the input $\psi(t)$ for any prescribed output $\gamma(t)$. If we define

$$b^{-1}(t) = \delta(t) + k\kappa\, e^{-\kappa t} u(t)$$

then we may write

$$\psi(t) = b^{-1}(t) * \gamma(t).$$

The laplace transform of $b^{-1}(t)$ is

$$B^{-1}(p) = 1 + k\kappa \frac{1}{p+\kappa} = \frac{p+\kappa+k\kappa}{p+\kappa}.$$

Hence

$$B(p) = \frac{p+\kappa}{p+\kappa+k\kappa} = 1 - \frac{k\kappa}{p+(1+k)\kappa}$$

and so

$$b(t) = \delta(t) - k\kappa\, e^{-(1+k)\kappa t} u(t).$$

We therefore have

$$\gamma(t) = b(t) * \psi(t) = \psi(t) - k\kappa \int_0^\infty e^{-(1+k)\kappa s} \psi(t-s)\, ds,$$

a linear equation which allows us to calculate the output $\gamma(t)$ for any prescribed input $\psi(t)$.

Neither the impulse response function $b(t)$ nor its inverse $b^{-1}(t)$ is dissipative, but both $b(t)$ and its inverse $b^{-1}(t)$ are realizable. We note that if κ is small then the AVC action is slow, and that if κ is large then the AVC action is fast.

Example 12.20. Let the input envelope be the damped exponential

$$f(t) = f_0 e^{\alpha - \beta t}, \quad \text{for} \quad t > 0, \quad (\beta > 0).$$

Then $\psi(t)$ is the straight line

$$\psi(t) = \log \frac{f(t)}{f_0} = \alpha - \beta t, \quad t > 0.$$

We have

$$\gamma(t) = b(t) * \psi(t)$$
$$= \frac{1}{1+k}\left\{\alpha - \frac{k\beta}{\kappa(1+k)} - \beta t + \left[k\alpha + \frac{k\beta}{\kappa(1+k)}\right] e^{-(1+k)\kappa t}\right\}.$$

Letting $\alpha = 1$, $\beta = 1$, $f_0 = 1$, $k = 1$, and $\kappa = 10$, we obtain the following table:

Time t	0·0	0·1	0·2	0·3	0·4	0·5	1·0	2·0
Input envelope f(t)	2·72	2·46	2·23	2·01	1·82	1·65	1·00	0·37
Output envelope g(t)	2·72	1·65	1·48	1·39	1·32	1·26	0·98	0·59

We see that the curves for $f(t)$ and $g(t)$ cross just beyond the point $t = 1{\cdot}0$. For values of t to the left of this point, the AVC system reduces the output envelope $g(t)$ in relation to the input envelope $f(t)$; and for values of t to the right of this point, the AVC system increases the output envelope in relation to the input envelope. That is, the AVC compensates for the exponential decay of input energy with time by attempting to maintain the output energy more nearly at a constant level through time.

REFERENCES

ACHIESER, N. I. (1956). *Theory of Approximation.* Ungar, N.Y.

AKUTOWICZ, E. J. (1959). On extrapolating a positive definite function from a finite interval. *Math. Scand.*, **7**, 157–169.

BALABANIAN, N. and LEPAGE, W. (1956). What is a minimum-phase network ? *Amer. Inst. Elect. Eng.*, 785–787.

BALAKRISHNAN, A. V. (1960). Estimation and detection theory for multiple stochastic processes. *J. Math. Analysis and Applications*, **1**, 386–410.

BARTLETT, M. S. (1955). *Stochastic Processes.* Cambridge Univ. Press.

BEER, S. (1959). *Cybernetics and Management.* English Univ. Press.

BELLMAN, R. (1961). *Adaptive Control Processes.* Princeton Univ. Press.

BOCHNER, S. (1955). *Harmonic Analysis and the Theory of Probability.* Univ. of Cal. Press, Berkeley.

BOCHNER, S. and CHANDRASEKHARAN, K. (1949). *Fourier transforms.* (*Ann. Math. Studies*, **19**.) Princeton Univ. Press.

BODE, H. W. (1945). *Network Analysis and Feedback Amplifier Design.* Van Nostrand, Princeton, N.J.

BODE, H. W. and SHANNON, C. E. (1950). A simplified derivation of linear least square smoothing and prediction theory. *Proc. Inst. Rad. Eng.*, **38**, 417–425.

BRILLOUIN, L. (1956). *Science and Information Theory.* Acad. Press, N.Y.

CHOVER, J. (1958). Conditions on the realization of prediction by measures. *Duke Math. Jour.*, **25**, 305–310.

CRAMÉR, H. (1940). On the theory of stationary random processes. *Ann. Math.*, **41**, 215–230.

CRAMÉR, H. (1942). On harmonic analysis in certain functional spaces. *Ark. Mat. Astr. Fys.*, **28**B, No. 12, 1–7.

CRAMÉR, H. (1959). On the linear prediction problem for certain stochastic processes. *Arkiv för Matematik*, **4**, 45–53.

CRAMÉR, H. (1961). On the structure of purely non-deterministic stochastic processes. *Arkiv för Matematik*, **4**, 249–266.

DAVENPORT, W. B. and ROOT, W. L. (1958). *An Introduction to the Theory of Random Signals and Noise.* McGraw-Hill, N.Y.

DOOB, J. L. (1953). *Stochastic Processes.* Wiley, N.Y.

DURBIN, J. (1960). Estimation of parameters in time-series regression models. *J. Roy. Statist. Soc.*, **B**, **15**, 139–153.

FURSTENBERG, H. (1960). *Stationary Processes and Prediction Theory.* (*Ann. Math. Studies*, **44**.) Princeton Univ. Press.

GELFAND, I. M. and SCHILOV, G. E. (1960). *Verallgemeinerte Funktionen* (*Distributionen*): Volume 1. VEB Deut. Verlag der Wissenschaften, Berlin.

GOLDBERG, R. R. (1961). *Fourier Transforms.* Cambridge Tracts in Math. and Math. Physics, **52**. Cambridge Univ. Press.

GRENANDER, U. (1950). Stochastic processes and statistical inference. *Arkiv för Matematik*, **1**, 195–277.

GRENANDER, U. and ROSENBLATT, M. (1957). *Statistical Analysis of Stationary Time Series.* Wiley, N.Y.

GRENANDER, U. and SZEGÖ, G. (1958). *Toeplitz Forms and their Applications.* Univ. of Cal. Press. Berkeley.

HANNAN, E. J. (1960). *Time Series Analysis.* Methuen, London.

HANNER, O. (1949). Deterministic and non-deterministic stationary random processes. *Arkiv för Matematik*, **1**, 161–177.

HELSON, H. and LOWDENSLAGER, D. (1958). Prediction theory and Fourier series in several variables. *Acta Math.*, **99**, 165–202.

HELSTROM, C. W. (1960). *Statistical Theory of Signal Detection.* Pergamon, London.

HIRSCHMAN, I. I. and WIDDER, D. V. (1955). *The Convolution Transform.* Princeton Univ. Press.

JENSEN, J. L. W. V. (1906). Sur les fonctions convexes et les inéqualités entre les valeurs moyennes. *Acta Math.*, **30**, 175–193.

KARHUNEN, K. (1947). Über lineare Methoden in der Wahrscheinlichkeitsrechnung. *Ann. Acad. Sci. Fennicae Ser. A, I. Math. Phys.* **37**, 1–79. (Spanish translation (1952): *Trabajos de Estadistica*, **3**, 59–137.)

KARHUNEN, K. (1949). Über die Struktur stationärer zufälliger Funktionen. *Arkiv för Matematik*, **1**, 141–160.

KENDALL, M. G. (1945). On the analysis of oscillatory time-series. *J. Roy. Statist. Soc.*, **108**, 93–129.

KENDALL, M. G. (1946). *Contributions to the Study of Oscillatory Time-Series.* Cambridge Univ. Press.

KHINTCHINE, A. YA. (1934). Korrelationstheorie der stationären stochastischen Prozesse. *Math. Ann.*, **109**, 604–615.

KOLMOGOROV, A. N. (1941). Stationary sequences in Hilbert space (Russian). *Bull. Moscow State Univ., Math.*, **2**, 40 pp. (Spanish translation (1953): *Trabajos de Estadistica*, **4**, 55–73 and 243–270.)

KREIN, M. G. (1945). On a problem of extrapolation of A. N. Kolmogorov (Russian). *C.R. (Doklady) Acad. Sci. U.R.S.S.* (N.S.), **46**, 306–309.

KREIN, M. G. (1954). On a fundamental approximation problem in the theory of extrapolation and filtering of stationary random processes (Russian). *Doklady Akad. Nauk SSSR* (N.S.), **94**, 13–16.

KRYLOV, V. I. (1939). On functions regular in a half-plane (Russian with German summary). *Mat. Sbornik* (N.S.), **6** (**48**), 95–138.

LANING, J. H. and BATTIN, R. H. (1956). *Random Processes in Automatic Control.* McGraw-Hill, N.Y.

LEE, Y. W. (1960). *Statistical Theory of Communication.* Wiley, N.Y.

LÉVY, P. (1948). *Processus Stochastiques et Mouvement Brownien.* Gauthier-Villars, Paris.

MASANI, P. (1959). Cramér's theorem on monotone matrix-valued functions and the Wold decomposition. *Probability and Statistics* (*The Harald Cramér volume*), Almqvist and Wiksells, Stockholm, and Wiley, N.Y., 175–189.

MASANI, P. (1960). The prediction theory of multivariate stochastic processes, III. *Acta Math.*, **104**, 141–162.

MASANI, P. and WIENER, N. (1959). Non-linear prediction. *Probability and Statistics* (*The Harald Cramér volume*), Almqvist and Wiksells, Stockholm, and Wiley, N.Y., 190–212.

MASANI, P. and WIENER, N. (1959a). On bivariate stationary processes and the factorization of matrix-valued functions. *Theory of Probability and its Applications*, **4**, 300–308.

MIDDLETON, D. (1960). *An Introduction to Statistical Communication Theory.* McGraw-Hill, N.Y.

MIKUSINSKI, J. (1959). *Operational Calculus* (5th edn). Pergamon, London.

NAGABHUSHANAM, K. (1951). The primary process of a smoothing relation. *Arkiv för Matematik*, **1**, 421–488.

PALEY, R. E. A. C. and WIENER, N. (1934). *Fourier Transforms in the Complex Domain.* Amer. Math. Soc. Colloquium Publication, **19**, Providence, R.I.

PARZEN, E. (1960). *A Survey of Time Series Analysis.* ONR Technical Report, **37**, Stanford Univ., Stanford, Cal.

PRIVALOV, I. I. (1950). *Boundary Properties of Analytic Functions* (Russian). Gostehizdat, Moscow–Leningrad.

QUENOUILLE, M. H. (1957). *The Analysis of Multiple Time-Series.* Griffin's Statistical Monographs and Courses **1**. Charles Griffin & Co., London, and Stechert-Hafner, N.Y.

ROBINSON, E. A. (1954). *Predictive Decomposition of Time Series with Applications to Seismic Exploration.* Ph.D. Thesis, M.I.T.

ROBINSON, E. A. (1957). Predictive decomposition of seismic traces. *Geophysics*, **22**, 767–778.

ROBINSON, E. A. (1959). *An Introduction to Infinitely Many Variates.* Griffin's Statistical Monographs and Courses **6**, Charles Griffin & Co., London, and Stechert-Hafner, N.Y.

ROBINSON, E. A. (1959a). A stochastic diffusion theory of price. *Econometrica*, **27**, 679–684.

ROBINSON, E. A. (1960). Sums of stationary random variables. *Proceedings Amer. Math. Soc.*, **11**, 77–79.

ROBINSON, E. A. (1962). Extremal representation of stationary stochastic processes. *Arkiv för Matematik*, **4**, 379–384.

ROBINSON, E. A., SIMPSON, S. M., and SMITH, M. K. (1954). *Further Research on Linear Operators in Seismic Analysis.* M.I.T. Geophysical Analysis Group, **6**, Cambridge, Mass.

ROSENBLATT, M. (1957). A multi-dimensional prediction problem. *Arkiv för Matematik*, **3**, 407–424.

SCHULTZ, W. C. and RIDEOUT, V. C. (1957). A general criterion for servo performance. *Proc. National Electronics Conference (Chicago)*, **13**, 1–12.

SCHWARTZ, L. (1950, 1951). *Théorie des Distributions.* Hermann, Paris.

SIMPSON, S. M. (1956). *Properties, Origin, and Treatment of Certain Types of Seismic Noise.* M.I.T. Geophysical Analysis Group, **10**, Cambridge, Mass.

SIMPSON, S. M. (1957). *The Interrelation of the Deterministic and Probabilistic Approaches to Seismic Problems.* M.I.T. Geophysical Analysis Group, **11**, Cambridge, Mass.

SMITH, M. K. (1959). Personal communication.

SOLODOVNIKOV, V. V. (1952). *Introduction to the Statistical Dynamics of Automatic Control Systems* (Russian). English translation (1960). Dover, N.Y.

STORER, J. E. (1957). *Passive Network Synthesis.* McGraw-Hill, N.Y.

SUNDSTRÖM, M. (1951). Some statistical problems in the theory of servomechanisms. *Arkiv för Matematik*, **2**, 139–246.

TITCHMARSH, E. C. (1948). *Introduction to the Theory of Fourier Integrals.* Oxford Univ. Press.

TRUXAL, J. G. (1955). *Automatic Feedback Control System Synthesis.* McGraw-Hill, N.Y.

TSIEN, H. S. (1954). *Engineering Cybernetics.* McGraw-Hill, N.Y.

VAN DER POL, B. and BREMMER, H. (1950). *Operational Calculus.* Cambridge Univ. Press.

WADSWORTH, G. P. and BRYAN, J. G. (1960). *Introduction to Probability and Random Variables.* McGraw-Hill, N.Y.

WADSWORTH, G. P., ROBINSON, E. A., BRYAN, J. G. and HURLEY, P. M. (1953). Detection of reflections on seismic records by linear operators. *Geophysics*, **18**, 539–586.

WHITTLE, P. (1953). The analysis of multiple stationary time series. *J. Roy. Statist. Soc.*, **B**, **15**, 125–139.

WIDDER, D. V. (1941). *The Laplace Transform.* Princeton Univ. Press.

WIENER, N. (1933). *The Fourier Integral.* Dover, N.Y.

WIENER, N. (1942). *Stationary Time Series.* M.I.T. N.D.R.C. Report, Cambridge, Mass. (Reprinted, Wiley, N.Y.)

WIENER, N. (1958). *Nonlinear Problems in Random Theory.* Wiley, N.Y.

WIENER, N. and MASANI, P. (1957, 1958). The prediction theory of multivariate stochastic processes. *Acta Math.*, **98**, 111–150; **99**, 93–137.

WILCOX, C. (1959). Personal communication.

WOLD, H. (1938). *A Study in the Analysis of Stationary Time-Series.* Thesis, University of Stockholm (2nd edn, Almqvist and Wiksells, Uppsala, 1954).

WOLD, H. (1961). Unbiased predictors. *Proceedings Fourth Berkeley Symposium on Math. Statist. and Prob.* Univ. of Cal. Press, **1**, 719–761.

WOLD, H., in association with JURÉEN, L. (1953). *Demand Analysis.* Wiley, N.Y.

YAGLOM, A. M. (1952). Introduction to the theory of stationary random functions (Russian). *Uspekhi Mat. Nauk.* N.S. **7**, no. 5 (**51**), 3–168. (English translation by D. V. Thampuran, 1962, Interscience Press, N.Y.

YAGLOM, A. M. (1955). Extrapolation, interpolation and filtering of stationary random processes with rational spectral density (Russian). *Trudy Moskov. Mat. Obsc.*, **4**, 333–374.

YAGLOM, A. M. (1955a). Correlation theory of processes with random stationary *n*th increments (Russian). *Mat. Sbornik*, N.S. **37** (**79**), 141–196. (English translation (1958): *Amer. Math. Soc. Translations*, Series 2, **8**, 87–141.)

YULE, G. U. (1927). On a model of investigating periodicities in disturbed series with special reference to Wolfer's sunspot numbers. *Phil. Trans.*, **A**, **226**, 267–298.

ZASUHIN, V. (1941). On the theory of multidimensional stationary random processes. *C.R. (Doklady) Acad. Sci. U.R.S.S.*, **33**, 435–438.

INDEX